T0306004

Polymer-Based Smart Materials — Processes, Properties and Application

MATERIALS RESEARCH SOCIETY
SYMPOSIUM PROCEEDINGS VOLUME 1134

Polymer-Based Smart Materials — Processes, Properties and Application

Symposium held December 2–5, 2008, Boston, Massachusetts, U.S.A.

EDITORS:

Siegfried Bauer
Johannes Kepler University
Linz, Austria

Zhongyang Cheng
Auburn University
Auburn, Alabama, U.S.A.

Debra A. Wrobleski
Los Alamos National Laboratory
Los Alamos, New Mexico, U.S.A.

Qiming Zhang
The Pennsylvania State University
University Park, Pennsylvania, U.S.A.

Materials Research Society
Warrendale, Pennsylvania

CAMBRIDGE
UNIVERSITY PRESS

University Printing House, Cambridge CB2 8BS, United Kingdom

One Liberty Plaza, 20th Floor, New York, NY 10006, USA

477 Williamstown Road, Port Melbourne, VIC 3207, Australia

314-321, 3rd Floor, Plot 3, Splendor Forum, Jasola District Centre, New Delhi - 110025, India

79 Anson Road, #06-04/06, Singapore 079906

Cambridge University Press is part of the University of Cambridge.

It furthers the University's mission by disseminating knowledge in the pursuit of education, learning and research at the highest international levels of excellence.

www.cambridge.org
Information on this title: www.cambridge.org/9781605111063

Materials Research Society
506 Keystone Drive, Warrendale, PA 15086
http://www.mrs.org

© Materials Research Society 2009

First published 2009
First paperback edition 2012

Single article reprints from this publication are available through University Microfilms Inc., 300 North Zeeb Road, Ann Arbor, MI 48106

CODEN: MRSPDH

A catalogue record for this publication is available from the British Library

ISBN 978-1-605-11106-3 Hardback
ISBN 978-1-107-40840-1 Paperback

CONTENTS

DEVICE APPLICATION II

POSTER SESSION:
NEW MATERIALS AND DEVICES

*Invited Paper

NEW MATERIALS AND CHARACTERIZATION III

POSTER SESSION

E-NSF AND NEW MATERIALS

PREFACE

Symposium BB, "Polymer-Based Smart Materials—Processes, Properties and Application," held December 2–5 at the 2008 MRS Fall Meeting in Boston, Massachusetts, has shown up with a wide range of new materials, characterization techniques, and applications in devices. Smart materials convert energy from one form into another form, especially into electrical energy. Smart polymer materials are very attractive due to their flexibility, low-cost production and easy processability. Polymers have been shown to exhibit better performance than inorganic materials, such as larger strain responses and energy density.

The symposium provided a platform for researchers from academia and industry interested in smart polymer materials. It stimulated discussions among physicists, chemists, and engineers working in the field of smart plastics. With a focus on materials development, characterization, processing, manufacturing, analysis, design and applications, this proceedings summarizes selected work presented at the symposium.

Siegfried Bauer
Zhongyang Cheng
Debra A. Wrobleski
Qiming Zhang

July 2009

MATERIALS RESEARCH SOCIETY SYMPOSIUM PROCEEDINGS

MATERIALS RESEARCH SOCIETY SYMPOSIUM PROCEEDINGS

Prior Materials Research Society Symposium Proceedings available by contacting Materials Research Society

New Materials and Characterization I

Mater. Res. Soc. Symp. Proc. Vol. 1134 © 2009 Materials Research Society   1134-BB01-06

Molecular Adsorption and Fragmentation of Bromoform on Polyvinylidene Fluoride With Trifluoroethylene

Carolina C. Ilie[1], Jie Xiao[2], and Peter A. Dowben[2]
[1]Physics, SUNY Oswego, 7060 Route 104, Oswego, New York 13126-3599, U.S.A.
[2]Physics, University of Nebraska at Lincoln, 116 Brace Lab., Lincoln, NE, U.S.A.

ABSTRACT

Bromoform absorption on crystalline polyvinylidene fluoride with 30% of trifluoroethylene, P(VDF-TrFE 70:30) was investigated by photoemission and inverse photoemission and found to be associative and reversible. As on other surfaces, bromoform decomposition does occur, under electron irradiation, and one possible mechanism is dissociative electron attachment.

INTRODUCTION

The adsorption of halocarbons on metal surface is an active area in surface science for more than two decades [1]. Despite the fact that the electronic structure of the haloforms is very similar [2], most of the emphasis has been on chloroform adsorption [3-8], although recently, bromoform [9,10] and chloroform [11,12] adsorption and photodecomposition on molecular films like ice have been investigated. While chloroform ($CHCl_3$) adsorption on calixarene molecular films has been studied [13], bromoform ($CHBr_3$) adsorption on the copolymer P(VDF-TrFE 70:30) provides a more compelling comparison [14] as this is a ferroelectric polymer as well as an excellent dielectric [15-17].

EXPERIMENTAL

Ultrathin ferroelectric thin films of copolymer 70% vinylidene fluoride with 30% of trifluoroethylene, P(VDF-TrFE 70:30) were fabricated by Langmuir-Blodgett (LB) deposition technique on freshly-cleaved highly ordered pyrolytic graphite (HOPG) substrates from the water subphase [16-17]. The 5 molecular layers thick films (25 Å) were annealed in ultrahigh vacuum at 110° C before and after each experiment for 30 minutes [18-21]. The Langmuir-Blodgett (LB) method produces highly crystalline films of P(VDF-TrFE) as is evident from the surface structure investigated by STM and experimental band structure mappings [18-21].The bromoform ($CHBr_3$) was admitted to the ultrahigh vacuum system through a standard leak valve and the exposures are denoted in Langmuirs (1 L = 10^{-6} torr.sec). The combined ultraviolet photoemission (UPS) and inverse photoemission (IPES) spectra were used to characterize the molecular orbital placement of both occupied and unoccupied orbitals of the polymer substrate and the bromoform adsorbate. In both photoemission and inverse photoemission measurements, the binding energies are referenced with respect to the Fermi edge of gold or tantalum, in contact with the sample surface. The UPS data taken in this work was done with a helium lamp He I at $hv = 21.2$ eV and a large PHI hemispherical analyzer with an angular acceptance of ±100 or more [14,19-21]. The IPES were obtained by using variable incident energy electrons while measuring the emitted photons at a fixed energy (9.7 eV) using a Geiger-Müller detector with an instrumental linewidth of ~ 400 meV, as described elsewhere [14,19-21].

RESULTS AND DISCUSSION

As seen in Figure 1, the highest occupied molecular orbital (HOMO) to lowest unoccupied molecular orbital (LUMO) gap for copolymer P(VDF-TrFE) is much smaller than expected from semi-empirical method NDO-PM₃ (neglect of differential diatomic overlap, parametric model number 3) model calculations based on Hartree-Fock formalism. These differences are due to intermolecular interactions within the PVDF-TrFE film (solid state effects) and band structure, with a large band dispersion evident in the unoccupied molecular orbitals [19]. Although PM₃ is a simplistic semiempirical calculation, density functional theory (DFT) is

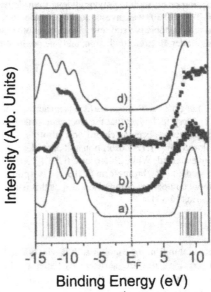

Figure 1: The combined UPS (left) and IPES (right) experimental spectra for clean PVDF-TrFE (b) and 30 L of bromoform adsorbed on PVDF-TrFE film at 120 K (c) compared to semiempirical models of the density of states for PVDF-TrFE (a) and bromoform (d). The molecular orbital eigen values are shown at the bottom and top respectively.

notorious for under estimating the band gap sometimes by a factor 2 or more [23], particularly for molecular systems, and must be rescaled for comparison with experiment [24], particularly final state spectroscopies like photoemission and inverse photoemission. To compare the model calculations with experiment (Figure 1), we applied Gaussian envelopes of 1 eV full width half maximum to each calculated molecular orbital energy (eigen value) to account for the solid state broadening in photoemission and then summing. These model density of states calculations are rigidly shifted in energy by 5.3 eV (representative of work function Φ equal to the difference of vacuum energy E_{vac} and Fermi level E_F) and then compared with the combined photoemission and inverse photoemission data, as indicated in Figure 1. No corrections are made for final state effects or matrix element effects in either calculation, so the comparison with experiment is simplistic, but still useful [14].

Following bromoform adsorption on PVDF-TrFE at 120 K, there is a clear suppression of the PVDF-TrFE photoemission features, and new features match the expected molecular orbitals of bromoform (Figure 1). This tends to suggest that bromoform adsorption "wets" the surface of PVDF-TrFE. The HOMO-LUMO gap, as determined from the combined photoemission and inverse photoemission, for bromoform adsorbed on PVDF-TrFE at 120 K is larger than expected based on the model semiempirical molecular calculations, as seen in Figure 1. This may be due to final state effects [24,25] that become more pronounced with decreasing temperature and the increased bromoform coverages. The possible decrease of photoemission final state screening is consistent with the fact that PVDF-TrFE is a good dielectric at 120 K. Thus bromoform adsorption on PVDF-TrFE at 120 K, should more closely resemble bromoform in the gas phase than the fully screened molecular adsorbate on a conducting substrate.

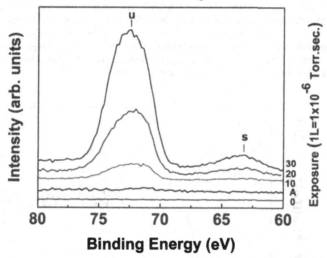

Figure 2: The sequence of photoemission spectra of a 5 monolayer thick P(VDF-TrFE 70:30) at 120 K film before and after CHBr$_3$ exposure (1 Langmuir (L)= 1x10^{-6} Torr.sec.). Spectrum A is after annealing to room temperature showing that molecular adsorption of bromoform on PVDF-TrFE is reversible.

Bromoform adsorption is observed in the X-ray photoemission of the bromine 3d core level as well. The intensity increases with increasing exposure, in as seen in Figure 2. The binding energy of 72.1±0.1 eV at low bromoform exposure increases to 72.6±0.1 eV at higher exposures to P(VDF-TrFE) at 120 K [14], as illustrated in Figure 2. In spite of our inability to resolve the 3d$_{5/2}$ to 3d$_{3/2}$ core levels, the bromine 3d XPS features exhibit larger binding energies than expected for the 3d$_{5/2}$ core level: a binding energy of 70.3 eV has been measured for condensed bromophenol blue [26], although the binding energy for gaseous bromoform is 76.8 eV [27]. This is consistent with the higher oxidation state of bromoform although final state effects (as mentioned above) could contribute to increased binding energies, as the polymer substrate is a nominal dielectric. The increase in core level binding energy with increasing coverage suggests a repulsive interaction between adsorbate species, so a stronger interaction with the substrate polymer is certainly possible, and indeed likely.

In spite of possible stronger interactions with the substrate than with adjacent adsorbate species, bromoform adsorption on P(VDF-TrFE) must be weak: chloromethane, with a similar static dipole, is not seen to molecularly adsorb on P(VDF-TrFE) at all at 120 K. The adsorbed species whose signature is evident in photoemission, inverse photoemission and XPS is molecular bromoform. Following bromoform adsorption, the core level XPS bromoform signal is no longer apparent at when the substrate is annealed to room temperature, as the adsorbate bromoform desorbs below room temperature (Figure 2).

The main bromine 3d core photoemission intensity increases with coverage, but after each annealing the bromine signal is lost, indicative of bromine desorption. Following adsorption at 120 K and photofragmentation (as in [9,10,28]), the bromine 3d core level signal is persistent to room temperature and above (120° C) (Figure 3) and we must infer that dissociative adsorption is not reversible and is persistent in XPS to well above room temperature, so we attribute the core level spectra in Figure 2 to molecular, not dissociative adsorption. The photodissociation of bromoform leads to bromine 3d XPS features at 71.4±0.2 and 75.1±0.3 binding energies: binding energies both larger and smaller than the molecular bromoform 3d feature at 72.6±0.1 eV binding energy (Figure 3.)

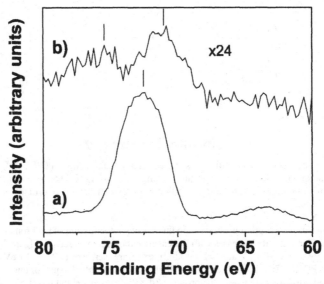

Figure 3: The photoemission spectra of photodissociated bromoform on P(VDF-TrFE) (top) leads to bromine 3d features at 71.4±0.2 and 75.1±0.3 binding energies: binding energies both larger and smaller than the molecular bromoform 3d feature at 72.6±0.1 eV binding energy.

This decompositon of bromoform is most efficient under electron irradiation, suggesting that a dissociative negative ion resonance may play a role. Such short lived negative ion states are known for a variety of bromoalkanes [29] and dissociative electron attachment has been identified for chlorofluormethanes in the gas phase [30-31]. While definitive identification of such dissociation mechanisms would require careful investigation of the decomposition rates as a function of electron kinetic energy, this remains a likely scenario.

CONCLUSIONS

We find evidence for molecular adsorption of bromoform on PVDF-TrFE. Chloromethane, with a similar static dipole, is not seen to molecularly adsorb on P(VDF-TrFE) at all at 120 K and we suspect it is a more weakly bound adsorbate compared to bromoform. As with other studies of bromoform adsorption, decomposition on an insulating substrate does seem facile under electron irradiation, even at low electron kinetic energies.

ACKNOWLEDGMENTS

We are grateful to Stephen Ducharme, Paul Burrow, Luis G. Rosa and R.G. Jones. This research was supported by National Science Foundation through grants CHE-0415421 and CHE-0650453.

REFERENCES

1. M. Grunze, P.A. Dowben, Appl. Surf. Sci. 10 209 (1982).
2. R. Tate, D.C. Driscoll, G. Z.Stauf, Naturforsch. 41a 1091 (1986).
3. R.G. Jones, C.A. Clifford, Phys. Chem. Chem. Phys. 1, 5223 (1999).
4. St. J. Dixon-Warren, D.V. Heyd, E.T. Jensen, J.C. Polanyi, J. Chem. Phys. 98, 5954 (1993).
5. St. J. Dixon-Warren, E.T. Jensen, J.C. Polanyi, J. Chem. Phys. 98, 5938 (1993).
6. R.G. Jones, I. Shuttleworth, C.J. Fisher, J.J. Lee, S.L. Bastow, R. Ithnin, J. Ludecke, M.P. Skegg, D.P. Woodruff, B.C.C. Cowie, Phys. Low-Dim. Structures 11-12, 1 (2001).
7. A. Bah, T. Ceva, B. Croset, N. Dupont-Pavlovsky, E. Ressouche, Surf. Sci. 395, 307 (1998).
8. J.M. Chen, S.C. Yang, Y.C. Lin, Surf. Sci. 391, 278 (1997).
9. M.L. Grecea, E.H.G. Backus, A.W. Kleyn, M. Bonn, Surf. Sci. 600, 3337 (2006).
10. M.L. Grecea, E.H.G. Backus, A.W. Kleyn, M. Bonn, J. Phys. Chem. B. 109, 17574 (2005).
11. J.E. Schaff, J.T. Roberts, Surf. Sci. 426, 384 (1999).
12. J.E. Schaff, J.T. Roberts, J. Phys. Chem. 100, 14151 (1996).
13. K.-D. Schierbaum, A. Gerlach, W. Gøopel, W.M. Müller, F. Vögtle, A. Dominik, H.J. Roth, Fresenius J. Anal. Chem. 349, 372 (1994).
14. J. Xiao, C. C. Ilie, N. Wu, K. Fukutani, P.A. Dowben, Surface Science, in press (2009).
15. T. Furukawa, Phase Transitions 18, 143 (1989).

16. L.M. Blinov, V.M. Fridkin, S.P. Palto, A.V. Bune, P.A. Dowben, S. Ducharme, Uspekhi Fizicheskikh Nauk [Russian edition vol.] **170**, 247 (2000); *Physics Uspekhi* [English edition volume] **43,** 243 (2000).

17. S. Ducharme, S.P. Palto, V.M. Fridkin, "Ferroelectric Polymer Langmuir-Blodgett Films", in: *Handbook of Surfaces and Interfaces of Materials*, vol. 3, *Ferroelectric and Dielectric Films*, chapter 11, (2002) p. 546-592.

18. H. Qu, W. Yao, T. Garcia, J. Zhang, S. Ducharme, P.A. Dowben, A.V. Sorokin, V.M. Fridkin, Appl. Phys. Lett. 82, 4322 (2003).

19. J. Choi, P.A. Dowben, S. Ducharme, V.M. Fridkin, S.P. Palto, N. Petukhova, S.G. Yudin, Physics Letters A 249, 505 (1998).

20. J. Choi, C.N. Borca, P.A. Dowben, A. Bune, M. Poulsen, S. Pebley, S. Adenwalla, S. Ducharme, Lee Robertson, V.M. Fridkin, S.P. Palto, N. Petukhova, S.G. Yudin, Phys. Rev. B 61, 5760 (2000).

21. Jie Xiao, L.G. Rosa, M. Poulsen, D.-Q. Feng, S. Reddy, J.M. Takacs, L. Cai, J. Zhang, S. Ducharme, P.A. Dowben, J. Phys. Cond. Matter 18, L155 (2006).

22. I.N. Yakovkin, P.A. Dowben, Surface Review and Letters 14, 481 (2007).

23. A.M. Scheer, P.D. Burrow, J. Phys. Chem. B 110, 17751 (2006).

24. J.E. Ortega, F.J. Himpsel, Dongqi Li, P.A. Dowben, Solid State Commun. 91, 807 (1994).

25. P.A. Dowben, Surface Science Reports 40, 151 (2000).

26. C.D. Wagner, J. Vac. Sci. Technol. 15, 518 (1978).

27. R. Spohr, T. Bergmark, N. Magnusson, L.O. Werme, C. Nordling, K. Siegbahn, Phys. Script. 2, 31 (1970).

28. W.S. McGivern, O. Sorkhabi, A.G. Suits, A. Derecskei-Kovacs, S.W. North, J. Phys. Chem. A 104, 10085 (2000).

29. S.A. Pshenichnyuk, N.L. Asfandiarov and P.D. Burrow, Russina Chem. Bulliten 56, 1268 (2007).

30. K. Aflatooni, and P.D. Burrow, Int. J. Mass Spectrosc. 205, 149 (2001).

31. S.C. Chu, and P.D. Burrow, Chem. Phys. Lett. 172, 17 (1990).

Mater. Res. Soc. Symp. Proc. Vol. 1134 © 2009 Materials Research Society   1134-BB01-09

Application of a Chemically Adsorbed Monolayer and Polypyrrole Thin Film for Increasing the Adhesion Force Between the Resin Substrate and the Plated Copper Layer

Yuji Ohkubo [1], Shogo Onishi [1], Satoshi Miyazawa [2], Kazuhiro Soejima [2] and Kazufumi Ogawa [1]

[1] Department of Advanced Materials Science, Graduate School of Engineering Kagawa University, Hayashi-cho, Takamatsu city, Kagawa 761-0396, Japan

[2] Business Development HQ Process Technology Dvlpmt. 15

Center, Alps Electric Co., Ltd. 3-31,
Akedori, Izumi-ku, Sendai city, Miyagi 981-3280, Japan

ABSTRACT

A chemically adsorbed monolayer containing pyrrolyl group (Pyrrolyl-CAM) was prepared between a plated copper layer and a resin substrate for increasing the adhesion force without roughening a surface of the resin substrate. Although it was not enough to increase the adhesion force between the resin substrate and the copper layer by using only Pyrrolyl-CAM, the sufficient adhesion force was obtained by preparing a polypyrrole thin film between Pyrrolyl-CAM and the copper layer.

Pyrrolyl-CAM and the polypyrrole thin film on the substrate were evaluated by an automatic contact angle meter and auger electron spectroscopy in order to analyze the condition of the films between the resin substrate and the copper layer.

The peel strength test was carried out in order to evaluate the adhesion force. The best adhesion force was 0.98 [N/mm], and the target value of 0.60 [N/mm] was sufficiently achieved.

INTRODUCTION

In recent years, high density integrated circuit, miniaturization of electronic components and improvement of printed-wire board play roles as high functionalizing and weight saving of electronic devices. The printed-wire board is made of a resin substrate and a copper film laminated, and patterned by an etching technique. Although the adhesion force between the copper film and the resin substrate is important in this process, the adhesion force between the organic and inorganic materials is not usually good.

Therefore, oxidative solutions are traditionally used for roughing the surface of the resin substrate to increase the anchor effect. But if the surface of the resin substrate is roughened, the clock frequency used will be decreased and making the copper wires with fine pitches will be difficult. In addition, the oxidative solutions, such as dichromic acid and permanganic acid, are harmful for the human and the earth's environment.

Consequently, a new technology has been studied for increasing the adhesion force between the copper layer and the resin substrate without roughening.

In this study, a chemically adsorbed monolayer (CAM) [1-4] containing pyrrolyl group and the polypyrrole thin film were introduced between the copper layer and the resin substrate for increasing the adhesion force without roughening the resin surface.

When the oxidative solutions are used for roughing the surface of the resin substrate in order to increase the anchoring effect, the values of the peel strength force were 0.3~0.6 [N/mm] in our test results. These values satisfied the safety and efficiency of the wiring. This is the reason why the target value of 0.60 [N/mm] was determined.

EXPERIMENTAL DETAILS

Materials

The resin substrates, 50 [mm] × 20 [mm] × 2 [mm] composed with the mixture of polyphenylene sulfide (PPS) and cylindrical glass filler φ = 10 [μm], h = 100 [μm] (PPS : glass = 60 [%] : 40 [%]), were supplied from Alps Electric Co., Ltd.

N-[11-(trichlorosilyl)undecyl] pyrrole (PNN), which has a pyrrolyl group at the molecular end, a trichlorosilyl group at the other end, and a hydrocarbon chain group at the middle portion, was synthesized and supplied by Shin-Etsu Chemical Co., Ltd. and used as the adsorbent molecule in order to prepare PNN-CAM on the resin substrate.

Procedure

The experimental procedure is summarized in Fig. 1. (i) The resin substrates were previously cleaned by the oxygen plasma treatment at 600 [W] for 60 [s] by using RMD-150 (UL-VAC, inc.) in order to obtain the organic contamination free substrate, to increase hydroxyl groups on the substrate surface and to activate the substrate surface [5]. (ii) PNN was adsorbed monomolecularly by immersing the resin substrate in the adsorption solution containing PNN (concentration: 5×10^{-3} [mol/L]) in a mixed solvent of the dehydrated chloroform of 33 [%] and the dimethyl-silicone of 67 [%] for 2 [h] at room temperature according to the recommendation by Ogawa et al [6, 7]. This procedure was carried out in a dry air atmosphere below the humidity of 10 [%]. (iii) The substrate was taken out from the solution, and washed with chloroform and acetone. (iv) The resin substrate covered with PNN-CAM was immersed in the acetonitrile solution containing the pyrrole monomer (concentration: 0.2 [M]) for 3 [min] at room temperature. (v) After being taken out from the acetonitrile solution containing the pyrrole monomer, the sample was dried naturally. (vi) The oxidative polymerization was carried out for 5 [min] at 30 [°C] by immersing the substrate in the aqueous solution containing ferric chloride or palladium chloride (concentration: 0.85×10^{-3} [mol/L]) [8, 9]. By the above two processes, the polypyrrole thin film connected to PNN-CAM was prepared. (vii) The copper electroless plating was carried out by using the electroless copper plating kit Thru-Cup PEA Ver. 3 (Uyemura & Co., Ltd) on the polypyrrole thin film in order to form an organo-copper complex. (viii) The galvanic electroplating of copper was also carried out in order to increase the thickness of the plated copper layer for evaluating the adhesion force. The positive plate and the negative plate were immersed in a sulfuric acid aqueous solution of 10 [wt%] for 3 [min] at room temperature. Moreover, those plates were immersed in the copper sulfate plating solution TFE 2001 (ADEKA CORPORATION), followed by applying an electricity of 0.02 [A/m^2] for 90 [min] at room temperature, and washing by the purified water. (ix) The sample plates were dried by air gun.

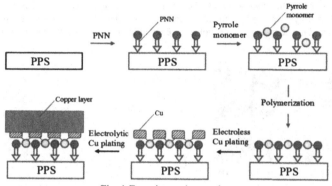

Fig. 1 Experimental procedure

Evaluation

The substrate was evaluated by using the automatic contact angle meter CA-VP150 (Kyowa Interface Science Co, LTD.), the auger electron spectrometer SAM 680 (ULVAC-PHI, INCOR-PORATED) and the tensile testing machine Strograph V10-β (Toyo Seiki Seisaku-sho, LTD.). The Peel strength force was calculated by following equation (1): where y is a loading, w is a width and n is a number of testing [10].

$$\text{Peel Strength force [N/mm]} = \frac{\left(\sum_{i=1}^{n} y_i\right)}{(n \cdot w)} \tag{1}$$

DISCUSSION

The changes of the water contact angles (WCA) with the processing on the resin substrate (PPS) are represented in Table 1. After the oxygen plasma treatment, the average water contact angles became ca. 3.4°. This result should indicate that hydroxyl groups are increased on the surface of the glass filler and that carboxyl, carbonyl and hydroxyl groups are formed on the surface of the oxidized resin substrate. After adsorbing PNN, the average water contact angles became ca. 71.3°, indicating that PNN-CAM should be prepared on the substrate surfaces. After oxidative polymerization, the average water contact angles became ca. 59.8°, indicating that the polypyrrole thin film should be prepared on the substrate surfaces covered with PNN-CAM.

Table.1 Changes of the water contact angles with the processing on the resin substrate (PPS)

Sample preparation condition	WCA [°]
Untreated substrate	86.6
After the oxygen plasma treatment	3.4
After the preparation of PNN-CAM	71.3
After the preparation of the polypyrrole thin film	59.8

11

Figure 2 shows the cross-sectional SEM image of the copper plated resin substrate after cleaning by the oxygen plasma treatment. It indicates that the interface between the resin substrate (PPS) and the copper layer was not roughened at the micron level by the oxygen plasma treatment.

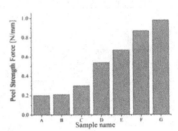

Fig. 2 Cross-sectional SEM image of the copper plated resin substrate after cleaning by the oxygen plasma treatment

Fig. 3 Adhesion force measured by peel strength test, indicating that adhesion force changes with the processing

Figure 3 shows the adhesion force changes with the processing. The force changes depended on some typical processes. The individual samples were prepared by the processes as follows. (A) : Sample A was untreated and used as a reference. (B) : Only PNN-CAM was prepared on the sample B. (C) : After preparing PNN-CAM on the sample C, it was immersed in the acetonitrile solution containing both the ferric chloride and the pyrrole monomer. (D) : After preparing PNN-CAM on the sample D, it was immersed in the acetonitrile solution containing only the pyrrole monomer, followed by immersing in the aqueous ferric chloride solution. (E) : After prepared PNN-CAM on the sample E, it was immersed in the acetonitrile solution containing only the pyrrole monomer, and then immersed in the aqueous ferric chloride solution, and the immersed again in the acetonitrile solution containing only the pyrrole monomer. (F) : The same process for the sample D was performed twice on the sample F. (G) : The same process as the sample F was performed on the sample G except only that the catalyst was changed from the ferric chloride to the palladium chloride.

Between sample A and B, there was a little difference of the adhesion forces, indicating that it was not much effective for the adhesion force to prepare only PNN-CAM between the plated copper layer and the resin substrate (PPS). Between sample B and C, the adhesion force increased almost 0.1 [N/mm] after the deposition of the polypyrrole thin film, indicating that it was effective to prepare the polypyrrole thin film on PNN-CAM for increasing the adhesive force.

Between sample C and E, the adhesion force increased almost 0.4 [N/mm] after the polymerization, indicating that the polymerization procedure is also important to increase the adhesion force. Between sample F and G, the adhesion force further increased and 0.98 [N/mm] was obtained on the Sample G. The value sufficiently achieved the target value of 0.60 [N/mm], indicating that changing the catalyst from the ferric trichloride to the palladium trichloride is effective to improve the adhesion force. These results may indicate that the ferric trichloride used for the oxidative polymerization might interrupt the deposition of the copper during the electroless copper plating.

Figure 4 is the auger electron spectrum on the resin substrate deposited by the same procedure as Sample G treated by the process of the PNN adsorption, oxidative polymerization, the electroless copper plating and the copper electrolytic plating. As expected, the atoms of copper, carbon, nitrogen, sulfur and oxygen were detected. In addition, this spectrum shows the bands due to the palladium and the chlorine, indicating that the residue of the catalyst remained on the sample.

Figure 5 is the depth profiles of copper, nitride, palladium, and sulfur atom (except the carbon atom) measured by using AES on the resin substrate deposited by the same procedure as Sample G. The black solid line containing square points shows the atomic concentrations of the copper in the plated copper layer. The black solid line containing triangular points shows the atomic concentration of the nitrogen that is a component of the polypyrrole thin film or PNN-CAM. The black dotted line containing rhombic points shows the atomic concentrations of the palladium that is the catalyst for the oxidative polymerization or for the electroless copper plating. The gray solid line containing cross points shows the atomic concentration of the sulfur that is a component of the resin substrate (PPS). The gray dotted line shows the atomic concentrations of the oxygen that is a component of PNN-CAM. The gray solid line containing stick points shows the atomic concentrations of the carbon that is a component of the polypyrrole thin film or PNN-CAM or resin substrate. The black solid line containing circular points shows the atomic concentrations of the chloride that is a residue of the catalyst.

At the beginning of sputtering, the atomic concentration of the copper was the highest in comparison with other atoms. The other atomic concentrations became less in the order of nitrogen and oxygen, indicating that the outer surface is the plated copper layer, and the polypyrrole thin film and PNN-CAM are deposited under the copper layer. With increasing the sputtering times, the atomic concentrations of the copper, nitrogen and oxygen were decreased, while the atomic concentration of the sulfur was increased, indicating that the bottom layer is the resin substrate (PPS). Additionally it indicates that PNN-CAM and the polypyrrole thin film are sandwiched between the plated copper layer and the resin substrate (PPS).

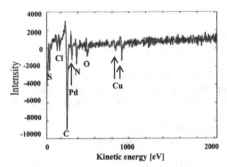

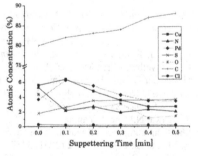

Fig. 4 Auger electron spectrum on the resin substrate deposited by the same procedure as Sample G

Fig. 5 Depth profile of copper, nitride, palladium, and sulfur by AES

CONCLUSIONS

The preparation of Pyrrolyl-CAM and the polypyrrole thin film on the substrate surface were confirmed by using the contact angle measurement. The existence of the copper layer, the polypyrrole thin film and Pyrrolyl-CAM were also confirmed on the resin substrate (PPS) by using AES.

The interface between the resin substrate (PPS) and the copper layer was observed by using SEM, and the cross-sectional SEM image indicated that the interface was not roughened at the micron level by the oxygen plasma treatment.

The polypyrrole thin film as the interlayer is effective for increasing the adhesion force between the plated copper and resin substrate (PPS). This may indicate that the organo-copper complex was formed at the interface between the polypyrrole thin film and the plated copper layer. The adhesion force between the plated copper and the resin substrate (PPS) increased to 0.98 [N/mm] by introducing the polypyrrole thin film and PNN-CAM between the plated copper layer and the resin substrate (PPS). The force value was about five times larger than that without those interlayers, and the demanded specification of the adhesive force 0.60 [N/mm] was fully achieved.

This technique should be useful for the preparation of the copper wires with fine pitches on the resin substrate without roughening.

ACKNOWLEDGMENTS

This study was partially supported by ALPS electronics Co., Ltd.. The authors thank Dr. Akira Nakano of ALPS electronics Co., Ltd. for his helpful discussions and coordinations.

REFERENCES

1. J.Sagiv: *J. Am. Chem. Soc.* **102** (1980) 92.
2. L. Netzer and J.Sagiv: *J. Am. Chem. Soc.* **105** (1983) 674.
3. K.Ogawa, N.Mino, H. Tamura and M. Hatada: *Langmuir* **6** (1990) 1807.
4. K.Ogawa, T.Ohtake, T.Nomura: *Jpn. J. Appl. Phys.* **39** (2000) 5904.
5. Uros˘ Cvelbar, Miran Mozetic˘, Ita Junkar, Alenka Vesel, Janez Kovac˘, Aleksander Drenik, Tjas˘a Vrlinic˘, Nina Hauptman, Marta Klanjs˘ek-Gunde, Bos˘tjan Markoli, Niks˘a Krstulovic´, Slobodan Milos˘evic´, Freddy Gaboriau, Thierry Belmonte: *Applied Surface Science* **253** (2007) 8669.
6. K.Ogawa, N.Mino, S.Yamamoto: *Thin Solid Films* **468** (2004) 240.
7. Shin-ichi Yamamoto, Kazufumi Ogawa: *Surface Science* **600** (2006) 4294.
8. A. Malinausksa: *polymer* **42** (2001) 3957.
9. M. Can, H. Ozaslan, O. Isldak: *polymer* **45** (2004) 7011.
10. JIS: HB series *Metal Surface Treatment* H8630 (Japanese Standards Association, Tokyo, 1998), p.272-273[in Japanese].

Mater. Res. Soc. Symp. Proc. Vol. 1134 © 2009 Materials Research Society 1134-BB01-10

Copying the Natural Skeletal Muscle Design Into a New Artificial Muscle System

Maria J. Bassil[1,2], Michael I. Ibrahim[1], Mario R. El Tahchi[1], Joseph K. Farah[1] and Joel Davenas[2]
[1] GBMI, Department of Physics, Lebanese University - Faculty of Sciences 2,
PO Box 90656 Jdeidet, Lebanon
[2] LMPB, Claude Bernard University -Lyon I, ISTIL Building, 15 Boulevard Latarjet,
69100, Villeurbanne, France

ABSTRACT

The key to make full use of gels is the design of the system in which they are used. So mimicking the natural skeletal muscle needs to create a new design that can regroup the electrical sensitivity, the linear displacement and the fast response while keeping good mechanical properties. Using a biomimetic approach, a new design of an artificial muscle based on hydrolyzed Polyacrylamide (PAAM) gel, is presented. The model consists on a fiber like elements of hydrolyzed PAAM, working in parallel, embedded in a thin conducting gel layer that plays the role of electrodes. The biocompatible electroactive microfibers are obtained by electrospinning method and the conductivity of three different gel made of PAAM-Polyanniline (PANI) blends is measured as a first step of developing gel electrodes.

INTRODUCTION

PAAM is an electroactif biocompatible non biodegradable [1-3] material. PAAM can not be absorbed into tissues and cells due to its high molecular weight which make it the most promising candidate for many bio-applications such as implants and artificial muscles [4-6]. Recently the electrochemical and the electromechanical properties, together with the actuation mechanisms of fully hydrolyzed PAAM hydrogel were studied [7-8] in order to obtain the elementary background needed for the design of actuating devices based on this material. Also the effect of geometrical dimensions of a cylindrical bulky gel on swelling and on the response time in a physiological solution was studied at the millimeter scale. It was assumed that microfibers with 500 μm diameter may contract to reach approximately 37% of its initial value in approximately 0.4s [9].

However the elaboration of such fibers has not been fully developed; in addition the structure-function relationship of PAAM based systems deserves further emphasis to demonstrate the profound influence of architecture on muscle function.

In this study a new design of an artificial muscle based on hydrolyzed PAAM gel is developed in order to be synthesized using an easy to process method. Based on the advantage of fibrous systems [10] and on the linear displacement actuation [9] this structure converts the electric energy into a linear displacement; it is able to receive electrical information and quickly transmit specific contraction responses. First, the model is presented, then the microfibers are obtained by electrospinning method, finally the conductivity of different PAAM-PANI blends is tested as a first step of developing gel electrodes.

THEORY

Model presentation

Based on the linear displacement actuator [9] and the principles of actuation mechanism of the gel [8]; the model consists on a fiber like elements of hydrolyzed PAAM, working in parallel, embedded in a thin conducting gel layer which plays the role of electrodes. Fibers are encapsulated in a braided structure by a thin elastic conductive gel film. The thin film holds the fibers together allowing them to act in parallel while keeping the response time as that of single fiber. This film distributes forces to minimize damage of the fibers and provides a conduit for electricity in order to stimulate the fibers. The film deposited in the longitudinal section of fibers forms the negative electrode while the film deposited in the cross-section of fibers forms the positive electrode. Under an electrical stimulation the fibers contract linearly. Like in natural muscle [11], as the stimulation increases, the number of fiber in state of contraction increases and the strength of contraction of the whole muscle also increases (figure 1 (d)).

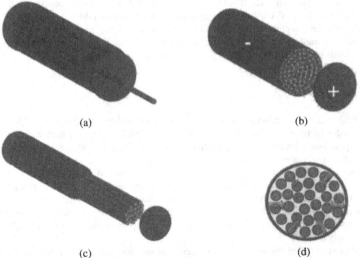

(a)

(b)

(c)

(d)

Figure 1. (a) (b) (c) Schematic representation of the artificial muscle system, (d) Cross section view.

Presentation of the ion exchange surface area

For the same volume we calculate the variation of the ion exchange surface area with respect to fibers radius. For modeling we consider L as the height of a bulky cylindrical gel of radius R. The ion exchange surface area of the cylinder is:

$$S_{(cylinder)} = 2\pi R^2 + 2\pi RL \tag{1}$$

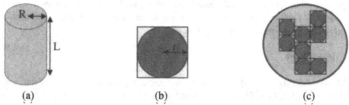

(a)   (b)   (c)

Figure 2. Schematic representation of (a) bulky cylindrical gel, (b) cross sectional view of gel fiber, (c) cross sectional view of gel fibers contained in the cylinder.

We assume that the cylinder is composed of fibers of radius $r < R$ and having the same height L. n is the number of fibers that can be contained in that volume. In addition, we assume that the area πR^2 can contain squares of side $2r$ (figure 2).

$$n = \frac{\pi R^2}{4r^2}$$

(2)

And the ion exchange surface of n fibers is:

$$S_{(n\,fibres)} = \frac{\pi^2 R^2}{2}(1 + \frac{L}{r}) \quad .$$

(3)

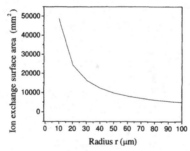

Figure 3. Ion exchange surface versus the radius of fibers for L=10cm, R=1mm.

Equation (3) is represented in figure 3. Fiber elements enable the system to exhibit relatively rapid response by providing a high surface area to volume ratio, which improves the flux rate of ions through the polymer matrix. A small diameter reduces the response lag, with respect to the applied stimulus, by enhancing the diffusion within the millimeter scale fibers themselves.

EXPERIMENT

Acrylamide 99%, N,N'-methylenebisacrylamide (BIS) ≥98% as cross-linker, N,N,N',N'-tertramethylethylenediamine 99% (TEMED), ammonium persulfate ≥98% (APS)

17

and aniline purum 99% was used. All chemicals were purchased from Aldrich and used as received without any further purification. Deionized water was used for all the dilutions, the polymerization reactions, as well as for the gel swelling.

Electrospun Microfibers

PAAM is synthesized by the standard free radical polymerization method using 1ml Acrylamide (30%), 60μl APS (25%), 20μl Temed with no crosslinker. This solution is used to control the viscosity of the precursor solution. After complete polymerization the resulting gel was diluted in 4ml of deionized water and 5ml Acrylamide (30%), 250μl BIS (2%), 10μl APS (25%) and 4μl Temed was added to form the precursor solution.

Fibers are obtained by the electrospinning technique. Figure 4 shows a schematic diagram of electrospinning process.

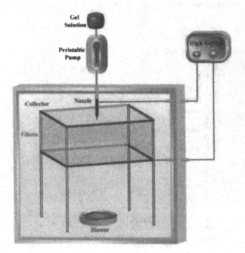

Figure 4. Schematic diagram of electrospinning process.

A plexiglas box of 60cm x 60 cm x 60 cm contains, a nozzle, a collector and a blower. The electric field is controlled by a high voltage power supply (Leybold 521721). The flow rate of the gel precursor is determined using the peristatic pump (Watson Marlow 101 U/R). The nozzle is stainless steel (I&J Fisnar) having a 0.6 mm inner diameter and 0.9 mm outer diameter. The nozzle is connected to the pump through a silicone tube of 0.5 mm inner diameter. The nozzle is glued on the upper plexiglas plate. Two aluminum square frames placed in parallel at 10 cm from each other form the collector. A copper wire connects these frames in order to have the same potential. Teflon rods fix the distance between the two frames; other Teflon rods separate these frames from the bottom of the box. The nose of a hot air gun is introduced at the bottom plate; it is situated around 30 cm from the nozzle. All the sides of the box where thermally insulated, and the air is evacuated continuously to maintain a fixed temperature inside the collector.

Conductive gel membrane

The two following solutions were prepared separately. Solution 1 to form polyacrylamide: 1ml Acrylamide (30%), 10μl APS (25%), 5μl Temed, 2μl BIS (2%) and solution 2 to form polyaniline (PANI): 1ml Aniline (25%), 300 μl APS(25%). After two minutes the solution 2 was poured slowly into the solution 1 in three different ratios (1ml PAAM / 1ml PANI, 1ml PAAM / 0.5ml PANI, 1ml PAAM / 0.2ml PANI). Each one of the resulting mixtures is placed on a microscope glass substrate to acquire complete polymerization. After complete drying the current-voltage characteristics of the samples are measured using a Keithley 487 pico-ammeter/voltage source.

DISCUSSION

The fibers are spun at 16 KV and grow in parallel between the aluminum frames (Figure 6 (a)). The viscosity of the precursor solution controls the fiber diameter. The distance between the aluminum frames can be adjusted to meet the required length of the fibers under certain limiting conditions. The obtained fibers are then thermally treated under 75°C for 12 hours. The resulting fibers are shown in figure 6 (b). The fiber swells rapidly when hydrated (Figure 6 (c)). The speed of swelling could not be measured due to it high value. The swelling ratio is about 10 times in diameter.

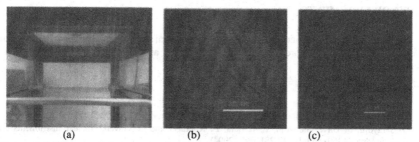

(a)   (b)   (c)

Figure 6. (a) Picture of electrospun microfibers, (b) optical image of fully swelled microfibers, (c) optical image of unswelled microfibers.

For the proposed model, formed by a system of 10cm in length and 1mm in radius; the elaboration of 10μm fibers can enhance the ion exchange surface 98 times (Figure 3).

The main task in artificial muscles development is based on the fact that volumetric changes occur in such polymers. Electrical stimulation is one of the most attractive stimulators that can produce elastic deformation in polymers. Since the volume variation of the gel structure is under the kinetic control of the driving electrical power; developing an adhesive, thin, flexible and conductive membrane which plays the role of electrodes can:
- Insure a large conduit for electricity in order to stimulate the gel while enhancing the response time.

- Holds the gel together and moves smoothly with the structure to enhance its mechanical properties.
As a first step of developing a conductive membrane the charge transfer conductivity in dried state of different PAAM-PANI gel blends is determined from current-voltage measurements. The charge transfer conductivity of the gel increase from 17.2×10^{-6} to 17×10^{-5} S.cm^{-1} while increasing the PANI concentration from 50% to 16.6%. This study should be developed thoroughly since we must be careful by using the appropriate amount of PANI in order to stay in the biocompatible range. Finding a compromise between conductivity and biocompatibility will be the next step before coupling the membrane to the fibers.

CONCLUSION

We cannot talk about polymer artificial muscle technologies, outside a system that possesses characteristics, resemble, or function like living organism. Based on the natural skeletal muscle design a new architecture muscle model is proposed and the first stage of the realization of the model is presented. Polyacrylamide microfibers are electrospun. These fibers show a fast swelling. Further studies have to be made in order to improve the model efficiency. This idea of copying, imitating, and learning from biology is the key to past, present, and future efforts in artificial muscle research.

ACKNOWLEDGMENTS

This work was carried out thanks to the financial support of the Lebanese National Council for Scientific Research, the Lebanese University and MIRA grant from Rhône Alpes region.

REFERENCES

1. E. Karadag, D. Saraydm, S. Cetinkayat and O. Giiven, Biomaterids 17, 67-70 (1996).
2. D. Saraydın, S. Unver-Saraydın, E. Karada, E. Koptagel and O. Guven, Nuclear Instruments and Methods in Physics Research B 217 281–292 (2004).
3. Jens Petersen, US patent No.7186419 B2, (6 Mars 2007).
4. Y.Osada and JP.Gong, Advanced Materials 10, 827-837 (1998).
5. Y.Bar-Cohen, in Electroactive *Polymer (EAP) Actuators as Artificial Muscles: Reality, Potential, and Challenges Second Edition*, edited by Y.Bar-Cohen, (SPIE Press Monograph, Bellingham, 2001)
6. N. Peppas, J. Zach Hilt, A. Khademhosseini, and Robert Langer, Adv. Mater. 18, 1345–1360 (2006).
7. M. Bassil, J. Davenas and M. El Tahchi, Sensors and Actuators B: Chemical 134, 496–501 (2008).
8. M. Bassil, M. El Tahchi, E. Souaid, J. Davenas, G. Azzi and R. Nabbout, Smart Materials and Structures 17, doi:10.1088/0964-1726/17/5/055017 (2008).
9. M. Bassil, M. El Tahchi and J. Davenas, Advances in Science and Technology 61, 85-90 (2008).
10. R. Bonser, W. Harwin, W. Hayes, G. Jeronimidis, G. Mitchell and C. Santulli, Report No. 18151/04/NL/MV, 2004.
11. B.MacIntosh, P.Gardiner and A.McComas, in *Skeletal Muscle: Form And Function 2 edition* (Human Kinetics Publishers, 2005)

Device Application I

Mater. Res. Soc. Symp. Proc. Vol. 1134 © 2009 Materials Research Society   1134-BB02-04

Composite Electromagnetic Wave Absorber Made of Aluminum Particles or Sendust Particles Dispersed in Polystyrene Medium

Kenji Sakai, Yoichi Wada, Yuuki Sato, and Shinzo Yoshikado
Department of Electronics, Doshisha University, 1-3 Tatara Miyakodani, Kyotanabe City, 610-0321, Japan

ABSTRACT

The frequency dependences of the relative complex permeability μ_r^*, the relative complex permittivity ε_r^* and the absorption characteristics for composite electromagnetic wave absorbers made of polystyrene resin and sendust (an alloy of Al, Si and Fe) or aluminum fine particles were investigated in the frequency range from 1 to 40 GHz. The size and volume mixture ratio of sendust and aluminum were varied. A metal-backed single-layer absorber made of sendust and that of aluminum absorbed more than 99% of electromagnetic wave power at frequencies above 10 GHz. Moreover, the values of μ_r^* was shown to be controlled by adjusting the volume mixture ratio and particle size of sendust or aluminum, and an electromagnetic wave absorber with a flexible design was proposed.

INTRODUCTION

The development of an electromagnetic wave absorber suitable for frequencies higher than 10 GHz is required with the increasing use of wireless telecommunication systems. To design a metal-backed single-layer absorber, the control of the frequency dependences of the relative complex permeability μ_r^* and the relative complex permittivity ε_r^* is important because the absorption of an electromagnetic wave is determined by μ_r^* and ε_r^*. Thus, the frequency dependences of μ_r^*, ε_r^* and the absorption characteristics for a composite made of a soft magnetic material dispersed in an insulating matrix have been investigated [1-3]. It has also been reported that μ_r' for a composite made of a soft magnetic material and an insulating matrix was less than unity [3]. This characteristic makes it possible to absorb 99% of the power of an electromagnetic wave at frequencies above 10 GHz because μ_r' must be less than unity to satisfy the absorption condition at these frequencies. Meanwhile, when an electromagnetic wave of high frequency enters a metal particle, such as an aluminum particle, an eddy current flows on the surface of a metal particle, and a reverse magnetic moment appears. This reverse magnetic moment results in the value of μ_r' of less than unity. Therefore, the absorption condition can be satisfied in the high-frequency range and the energy of the electromagnetic wave is converted into thermal energy by an eddy current. In addition, the material used for absorbers should be low-cost and available in large quantities to avoid the problem of worldwide resource depletion. Sendust and aluminum are low-cost materials because they do not contain any rare metals, hence these materials are suitable for use in an absorber.

In this study, the frequency dependences of μ_r^*, ε_r^* and the absorption characteristics for a composite made of sendust particles or aluminum particles dispersed in polystyrene resin were evaluated in the frequency range from 1 to 40 GHz. The volume mixture ratio and particle size of sendust and aluminum were varied, and the effects of the volume mixture ratio and particle size on the the absorption characteristics were investigated. From these results, the mechanism of the frequency dependence of μ_r^* was discussed in this paper.

EXPERIMENT

Commercially available sendust (Al 5%, Si 10%, Fe 85%) particles and aluminum particles were used. The average particle sizes (diameters) of the aluminum were approximately 8 and 30 μm, and those of sendust were approximately 5, 10 and 20 μm. Chips of polystyrene resin were dissolved in acetone. The dissolved polystyrene resin and sendust or aluminum particles were mixed. The volume mixture ratio V of sendust was 53 vol% (5, 10 and 20 μm) and 85 vol% (5 μm), and that of aluminum was 50 vol% (30 μm) and 16.4, 50, 60 vol% (8 μm). After mixing, the mixture was heated to melt the polystyrene resin and was then hot-pressed at a pressure of 5 MPa into a pellet. Then, the pellet was cooled naturally to room temperature and processed into a toroidal-core shape (with an outer diameter of approximately 7 mm and an inner diameter of approximately 3 mm) for use in a 7 mm coaxial line in the frequency range from 1 to 12.4 GHz, or into a rectangular shape for use in a waveguide in the frequency range from 12.4 to 40 GHz. The sample was loaded into the coaxial line or rectangular waveguide while ensuring that there was no gap between the walls of the coaxial line or the rectangular waveguide and the processed sample. The complex scattering matrix elements S_{11}^{*} (reflection coefficient) and S_{21}^{*} (transmission coefficient) for the TEM mode (coaxial line) or TE_{10} mode (rectangular waveguide) were measured using a vector network analyzer (Agilent Technology, 8722ES) by the full-two-port method. The values of μ_r^{*} ($\mu_r^{*} = \mu_r' - j\mu_r''$, $j = \sqrt{-1}$) and ε_r^{*} ($\varepsilon_r^{*} = \varepsilon_r' - j\varepsilon_r''$) were calculated from the data of both S_{11}^{*} and S_{21}^{*}. The return loss R for each sample thickness was calculated from the complex reflection coefficient Γ^{*} using the relation $R = 20 \log_{10}|\Gamma^{*}|$.

RESULTS and DISCUSSION

Frequency dependence of μ_r^{*} for the composite made of sendust particles

Figure 1 shows the frequency dependences of μ_r' and μ_r'' for composites made of sendust particles and polystyrene resin. μ_r' for all composites made of sendust decreased with increasing frequency, was minimum in the frequency range from 12 to 20 GHz, then increased with increasing frequency. The minimum value of μ_r' for the composites made of 85 vol% sendust was low compared with other three composites. Meanwhile, the frequency at which μ_r' was minimum was independent of both V and particle size of sendust. When V of sendust is the same, the frequency dependence of μ_r' did not depended on the particle size at frequencies below 20 GHz. However, they were different with increasing frequency. μ_r' was largest for the composite made of sendust particles of 5 μm diameter and decreased in the order of particles with 10 and 20 μm diameter (hereafter, the word "diameter" is omitted) at frequencies above 28 GHz. μ_r'' for all composites were almost the same at frequencies below 20 GHz in spite of the difference of V and particle size. However, the frequency dependence of μ_r'' depended on V and the particle size of sendust in the high-frequency range. μ_r'' for the composite made of sendust particles of 10 and 20 μm increased with increasing frequency although μ_r'' for the composite made of sendust particles of 5 μm was almost zero. It is speculated that the particle size dependence of μ_r'' in the high-frequency range correlates with that of μ_r'. Meanwhile, μ_r'' for the composite made of 85 vol% sendust increased with increasing frequency. These results suggest that the frequency dependences of μ_r' and μ_r'' can be controlled by V and particle size. To investigate the difference of μ_r^{*}, the mechanism of the frequency dependence of μ_r^{*} for the composites made of sendust is discussed.

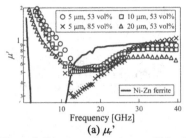
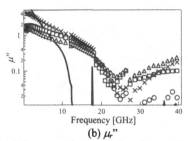

| O 5 μm, 53 vol% | □ 10 μm, 53 vol% |
| X 5 μm, 85 vol% | Δ 20 μm, 53 vol% |

— Ni-Zn ferrite

(a) μ_r' (b) μ_r''

Figure 1. Frequency dependences of (a) μ_r' and (b) μ_r'' for composites made of sendust particles and polystyrene resin. Lines show the frequency dependences of μ_r' and μ_r'' for the Ni-Zn ferrite.

One reason for the mechanism of the frequency dependence is estimated to be the natural magnetic resonance of sendust, because sendust is a magnetic material such as ferrite. To confirm the effect of magnetic resonance on μ_r^*, the frequency dependence of μ_r^* for the Ni-Zn ferrite is shown in figure 1. μ_r' for the ferrite decreased in the low-frequency range, was minimum, then increased with increasing frequency. μ_r'' for the ferrite was maximum in the low-frequency range and decreased with increasing frequency. This frequency dependence of μ_r^* was similar to that for the composite made of sendust. Therefore, the frequency dependence of μ_r^* for the composite made of sendust is speculated to be the natural magnetic resonance in the low-frequency range. However, the frequency dependence of μ_r^* for the ferrite and that for composites made of sendust were different in the high-frequency range and particle size dependence of μ_r^* can not be explained by the natural magnetic resonance. Then, another reason which is the generation of magnetic moments by the eddy current flowing on the surface of sendust particles can be considered, because sendust is conductive. To investigate this phenomenon, the frequency dependence of μ_r^* for the composite made of aluminum and the polystyrene resin is discussed in the next section.

Frequency dependence of μ_r^* for the composite made of aluminum particles and the mechanism of the frequency dependence of μ_r^*

The frequency dependences of μ_r' and μ_r'' for the composites made of aluminum particles and polystyrene resin are shown in figure 2. μ_r' for all composites made of aluminum was less than unity. This result is explained as follows. The skin depth δ of aluminum, which is from 2.6 to 0.4 μm in the frequency range from 1 to 40 GHz, is less than the diameter (approximately 8 and 30 μm) of the aluminum particles. Therefore, the eddy current flows in the layer of thickness δ. The eddy current generates a magnetic moment antiparallel to the incident magnetic field. Consequently, μ_r' is reduced and becomes less than unity. In addition, μ_r' decreased and μ_r'' increased with increasing V. These results were in approximate agreement with equation 1 obtained by a qualitative theoretical estimation [4].

$$\mu_r' = 1 - V, \qquad \mu_r'' = 2V\delta / a \qquad (1)$$

Here, a is the radius of aluminum particle. The lines shown in figure 2 represent the values calculated from equation 1. μ_r'' for the composite made of aluminum particles of 30 μm was

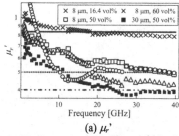
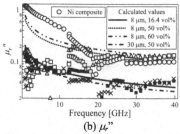

| (a) μ_r' | (b) μ_r'' |

Figure 2. Frequency dependences of (a) μ_r' and (b) μ_r'' for composites made of aluminum particles and polystyrene resin, and for the composite made of nickel particles and polystyrene resin.

lower than that for the composite made of aluminum particles of 8 μm. This result is in agreement with equation 1 and the particle size dependence is also explained by equation 1. However, μ_r' for all composites decreased gradually with increasing frequency and μ_r' for the composite made of aluminum particles of 30 μm was lower than that for the composite made of aluminum particles of 8 μm. These results are different from equation 1. The reason for this difference is speculated to be that the model used in calculation is ideal. The magnitude of magnetic moment is given by the product of the eddy current and the area S inside which the eddy current flows. If the frequency increases, S increases because the eddy current flows in the thin layer of particle surface. Moreover, S also increases with increasing the particle size. Thus, the magnitude of magnetic moment increases and μ_r' decreases as the frequency or particle size increases.

To clarify the effects of the magnetic resonance and the magnetic moment by the eddy current, μ_r' and μ_r'' for the composite made of nickel particles and the polystyrene resin were also measured and the measured values are shown in figure 2. Nickel is a magnetic material and is conductive. Therefore, the effects of both the magnetic resonance and the magnetic moments generated by the eddy current are observed in the composite made of nickel. The particle size (diameter) of nickel particles was between 10 and 20 μm and V of nickel was 50 vol%. In the low-frequency range, the frequency dependence of μ_r for the composite made of nickel was similar to that for the ferrite and to that for the composite made of sendust. Meanwhile, μ_r' for the composite made of nickel decreased at frequencies above 20 GHz and approached 0.5 although μ_r' for the ferrite increased with increasing frequency. This decrease resembles the decrease in μ_r' for the composite made of aluminum, and the value to which it approached almost agreed with that obtained from equation 1. In addition, μ_r'' for the composite made of nickel almost agreed with that for the composite made of aluminum at frequencies above 20 GHz although μ_r'' for the ferrite was almost zero. These results indicate that the effect of the magnetic moment by the eddy current is dominant in the high-frequency range. As shown in figure 1 (a), μ_r' for the composites made of sendust particles of 53 vol% were almost constant at frequencies above 28 GHz. This result indicates that the frequency dependence of μ_r' for the composite made of sendust in the high-frequency range is due to the magnetic moments by the eddy current. However, μ_r'' for the composites made of sendust particles of 10 and 20 μm increased with increasing frequency at frequencies above 26 GHz and became large as the particle size increased. These results are different from those obtained from equation 1. Therefore, another

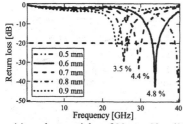 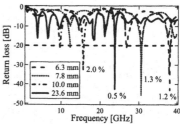

(a) sendust particles of 20 μm, 53 vol%   (b) aluminum particles of 8 μm, 50 vol%

Figure 3. Values of return loss for the composite made of sendust particles of 20 μm and that made of aluminum particles of 8 μm. The volume mixture ratio of sendust and aluminum is 53 vol% and 50 vol%, respectively.

reason for the frequency dependence of μ_r^* for the composite made of sendust must be considered in addition to the magnetic moments by the eddy current.

Absorption characteristics for the composite made of sendust or aluminum particles

The values of return loss for the sample thicknesses between 0.1 and 30 mm were calculated and good absorption characteristics were obtained for the composite made of 53 vol%-sendust particles of 20 μm and that made of 16.4 vol%-aluminum particles of 8 μm. The values of return loss for these two composites are shown in figure 3. The percentages shown in the graphs represent the normalized −20 dB bandwidth (the bandwidth Δf corresponding to the return loss of less than −20 dB divided by the absorption center frequency f_0). The value of −20 dB corresponds to the absorption of 99% of the electromagnetic wave power. The composite made of sendust particles of 20 μm had a return loss of less than −20 dB in the frequency range from 20 to 40 GHz. In particular, the normalized −20 dB bandwidth was broad, and the sample thickness for which the return loss was less than −20 dB was less than 1 mm. Meanwhile, the composite made of 16.4 vol%-aluminum particles of 8 μm exhibited a return loss of less than −20 dB at frequencies above 10 GHz although the normalized −20 dB bandwidth was narrow and the sample thicknesses were thick.

To investigate the difference of absorption characteristics between the composite made of sendust particles of 20 μm and that made of 16.4 vol%-aluminum particles of 8 μm, the measured values of μ_r^* and the calculated values of μ_r^* that satisfy the nonreflective condition given by equation 2 [5] are shown in figure 4.

$$\sqrt{\mu_r^*/\varepsilon_r^*}\ \tanh(\gamma_0 d\sqrt{\mu_r^*\varepsilon_r^*}) = 1 \qquad (2)$$

Here, γ_0 is the propagation constant in free space and d is the sample thickness. ε_r' used for calculation is independent of frequency and is the same as the measured value (composite made of sendust particles of 20 μm: $\varepsilon_r' = 19$, composite made of 16.4 vol%-aluminum particles of 8 μm: $\varepsilon_r' = 6$). ε_r'' is assumed to be zero. As shown in figure 4, the calculated values of both μ_r' and μ_r'' intersected the measured values for the composite made of sendust near 34 GHz at a sample thickness of 0.6 mm. Therefore, the absorption of a large amount of electromagnetic wave power occurred around 34 GHz, as shown in figure 3 (a). Meanwhile, the measured values of μ_r'' for the composite made of aluminum did not agree with the calculated values near 31 GHz

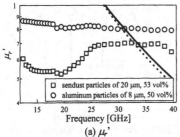 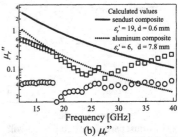

(a) μ_r' (b) μ_r''

Figure 4. Measured and calculated values of (a) μ_r' and (b) μ_r''. Plots show the measured values for the composite made of sendust particles of 20 μm and that of 16.4 vol%-aluminium particles of 8 μm. Lines show values calculated using equation 2.

although the measured values of μ_r' agreed with the calculated values near 31 GHz. Thus, the bandwidth for the composite made of 16.4 vol%-aluminum particles of 8 μm was narrow. However, as given by equation 1, μ_r'' can be controlled so that equation 2 is satisfied by adjusting the particle size of aluminum without changing μ_r'. This result suggests that the absorption with a wide bandwidth is possible for the composite made of aluminum particles near 31 GHz. In addition, the frequency at which the absorption occurs can be selected because μ_r' and μ_r'' can be controlled independently by adjusting V and particle size of aluminum. Therefore, a design of an absorber with good absorption characteristics at any frequency is expected.

CONCLUSIONS

The composite made of sendust particles and that made of aluminum particles exhibited a return loss of less than −20 dB at frequencies above 10 GHz. It was found that the frequency dependences of μ_r' and μ_r'' depended on the volume mixture ratio and the particle size of sendust and aluminum. In particular, μ_r' and μ_r'' for the composite made of aluminum particles almost agreed with the values calculated from the qualitative theoretical estimation. Therefore, an absorber with a flexible design was proposed since μ_r' and μ_r'' can be controlled artificially by the volume mixture ratio and particle size.

ACKNOWLEDGMENTS

This work was supported by "Kyoto Prefecture Collaboration of Regional Entities for the Advancement of Technological Excellence, JST", "the RCAST of Doshisha University and the Joint Research Project in the Kyoto Prefectural Area" and Nippon Atomized Metal Powders Corporation.

REFERENCES

1. K. M. Lim, K. A. Lee, M. C. Kim, and C. G. Park, *J. Non-Cryst. Solids.* **351**, 75 (2005).
2. L. Olmedo, G. Chateau, C. Deleuze, and J. L. Forveille, *J. Appl. Phys.* **73**, 6992 (1993).
3. T. Kasagi, T. Tsutaoka, and K. Hatakeyama, *Appl. Phys. Lett.* **88**, 17502 (2006).
4. Y. Wada, N. Asano, K. Sakai, and S. Yoshikado, *PIERS Online* **4**, 838 (2008)
5. Y. Naito and K. Suetake, *IEEE Trans. Microwave Theory Tech.*, **19**, 65 (1971).

Optimization of New Ultralow-k Materials for Advanced Interconnection

Xuan Li, Professor James Economy
University of Illinois at Urbana-Champaign, Department of Materials Science and Engineering,
Urbana, Illinois, 61801, U.S.A.

ABSTRACT

The demand for increased signal transmission speed and device density for the next generation of multilevel integrated circuits has placed stringent demands on materials performance. Currently, integration of the ultra low-k materials in dual Damascene requires chemical mechanical polishing (CMP) to planarize the copper. Unfortunately, none of the commercially proposed dielectric candidates display the desired mechanical and thermal properties for successful CMP. A new polydiacetylene thermosetting polymer (PDEB-TEB) which displays a low dielectric constant (low-k) of 2.7 was recently developed. This novel material appears to offer the only avenue for designing an ultra low k dielectric (1.85k), which can still display the desired modulus (7.7Gpa) and hardness (2.0Gpa) sufficient to withstand the process of CMP.

In this talk, we will present recent additional studies to further characterize the thermal properties of spin-on PDEB-TEB ultra-thin film. These include the coefficient of thermal expansion (CTE), biaxial thermal stress, and thermal conductivity. Thus the CTE is $2.0*10^{-5}K^{-1}$ in the perpendicular direction and $8.0*10^{-6}K^{-1}$ in the planar direction. The low CTE provides a better match to the Si substrate which minimizes interfacial stress and greatly enhances the reliability of the microprocessors. Initial experiments with oxygen plasma etching suggest a high probability of success for achieving vertical profiles.

INTRODUCTION

The demand for increased signal transmission speed and device density in the next generation of multilevel integrated circuits has placed stringent demands on materials performance. [1] The *International Technology Roadmap for Semiconductors (ITRS), 2005 Update* indicates that ultra-low k materials will be required for interlevel metal insulators. [2] Currently, integration of the ultra low-k materials in dual Damascene requires chemical mechanical polishing (CMP) to planarize the copper. Unfortunately, none of the proposed dielectric candidates display the desired mechanical and thermal properties for successful CMP.[3]

Recent advances in the synthesis of a new family of aromatic polymers made in our group permit for the first time the ability to design a dielectric with an ultra-low k value with excellent mechanical and thermal properties. Success in this endeavor will result in new market opportunities of at least 600 million dollars per year for low k dielectrics. [4]

Work in the past three years in our group has shown that a condensed aromatic polynuclear based on diethynyl and triethynyl benzene has an extremely high modulus and hardness [5, 6]. Such a system could provide a unique solution to the problem compared to any other known family of polymers. [7, 8] Chemical nature of the oligomer is shown below in Figure 1 with a proposed structure for the cured polymer. [9] As expected, such a polymer is very thermally stable with practically zero weight loss at 400°C. Curing of the oligomer was accompanied by a larger exotherm at 280°C. [10] Exotherm is not expected to be a problem since the dielectric thickness is 1.0μm and below, where excess heat would be easily dispersed. The oligomer is cured in the presence of porogen such as low M_w polystyrene or abietic acid to yield closed pores from 18 nm (polystyrene) to 5nm (abietic acid). In Table I are summarized the electrical properties of these films containing from zero to 40% porogen. [11] This new family of polymers displays a very high hardness and modulus, properties essential for successful CMP. As can be seen in Figure 2, the mechanical properties of samples containing up to 40% porogen are as good as practically all of the candidate materials with no porogen shown in Table 2. This means that we appear to have met the original goals of the ITRS roadmap with respect to achieving the desired low k material, with the requisite modulus and hardness.

The cured polymer with no porogen displays an outstanding thermal resistance (1.56% weight loss at 425°C after 8hours). An important goal was to show that use of porogens such as low M_W polystyrene yielded a closed micropore structure. This was successfully demonstrated using scanning electron microscopy. There is almost a linear relationship in a plot of dielectric constant with amount of porogen (Table I). Most importantly values as low as 1.85 could be obtained with 40% porogen, which is significantly lower than the goals of the ITRS roadmap.

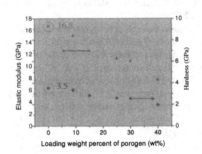

Figure1.Schematic synthesis of

DEB-co-TEB

Figure 2. Young's modulus and hardness

of 700nm film on Si

Table I. Electronic properties of cured DEB-co-TEB

Porogen (wt%)	Capacitance (pF)	Thickness (μm)	Dielectric Constant	Dielectric Loss	Breakdown Strength (V/μm)
0	6.2	0.85	2.70	0.002	> 235
9%	5.89	0.86	2.59	0.02	> 232
15%	5.28	0.90	2.43	0.04	> 222
20%	4.92	0.95	2.39	0.03	> 211
25%	4.47	1.02	2.26	0.02	> 196
30%	4.17	1.00	2.13	0.02	> 192
35%	-	0.92	2.03	0.03	> 217
40%	-	0.91	1.85	0.02	> 220

Porogen: Polystyrene

EXPERIMENTS
Materials

m-Diethynylbenzene monomer was purchased from Lancaster Synthesis Inc. (Pelham, NH), 1, 3, 5-Triethynylbenzene was obtained from Alfa Aesar (Ward Hill, MA). Polyimide (PI-2808) was purchased from HD MicroSystems (Parlin, NJ). Other chemicals were purchased from Aldrich (Milwaukee, WI). All chemicals were used as received except for abietic acid which was purified by heating to 260°C under vacuum twice and then only collecting the portion at 260°C.

Synthesis of Oligomer (m-Diethynylbenzene)-co-(Triethynylbenzene) (DEB-co-TEB)

14.0 g (0.11 mol) of m-diethynylbenzene (DEB), 5.56 g (0.037 mol) of 1,3,5-triethynylbenzene (TEB), extra phenylacetylene, and 8.3 mL of an acetone/pyridine mixture (50:50 volume ratio) in 200 mL of acetone was added into a vigorously stirred catalyst solution containing 13.3 g (0.13 mol) of Cu_2Cl_2, 8.3 mL of the acetone/pyridine mixture (50:50 volume ratio), and 700 mL of acetone. The catalyst solution was pre-oxidized by bubbling O_2 through for about 10 min before addition. The O_2 flow was continued throughout the rest of the reaction. 16.6 mL of the acetone/pyridine mixture was added over a period of 10 min. The reaction was conducted in the dark by covering the reaction flask with aluminum foil for 12 h. The reaction mixture was poured into HCl/methanol solution. The precipitate was collected, washed with methanol, and dissolved in chloroform. The solution was washed with HCl and deionized water, dried with $MgSO_4$, filtered, and precipitated from methanol. The solid was vacuum-dried at 50-55°C overnight to afford 10.2 g of a light-orange powder.

Fabrication of Dense Thin Film

A 25-35 wt % DEB-co-TEB oligomer in cyclohexanone solution was spin-coated onto a silicon wafer to form a thin film. The thin film was soft-baked at 110°C for 1 min in air, cured at 200°C for 30 min, and heated at 450°C for 30 min in a vacuum annealer.

Fabrication of Porous Film

Low molecular weight polystyrene (M_w = 780) was used as a porogen and dissolved in THF to provide a 25 wt % solution A. Another type of solution A was prepared by replacing polystyrene with abietic acid and selecting acetone as a solvent. Solution B was prepared by dissolving DEB-co-TEB into cyclohexanone at a 25 wt % concentration. Solutions A and B were mixed at various weight ratios to form a new solution by stirring at room temperature for 2 h and then allowing to sit for over 12 h. The resulting solutions were spin-coated onto Si wafer dices or metal (gold)-deposited Si wafer dices and soft-baked on a hot plate at 110°C for 1 min to remove most of the solvent. The wafer was then cured in a vacuum annealer under controlled heating. A typical curing recipe for forming a porous film was as follows: ramp at 5°C/min to 200°C, hold at 200°C for 30 min, ramp at 5°C/min to 250°C and hold for 30 min, ramp at 5°C/min to 350°C and hold for 30 min, and then ramp at 5°C/min to 450°C and hold for 30 min to burn out the porogen. The film was then cooled to room temperature at 5°C/min.

Thermal Properties Characterization

Before thermal measurements, the surface of the diffusion multiple was cleaned with polishing cloth to remove the trace of the microprobe and then coated with an Al film of 116 nm in thickness by magnetron sputtering at room temperature. The thermal conductivity is measured by time-domain thermoreflectance.

RESULTS

The research at its present stage of development appears to offer the avenue for designing an ultra low k dielectric, which can display the desired modulus and hardness values to withstand the process of CMP. Initial experiments with oxygen etching suggest a high probability of success for achieving vertical profiles (Figure3). This system can potentially be implemented directly into an assembly line for spin coating of dielectric followed by porogen evolution to yield the desired dielectric. The near and long term significance to the semiconductor industry would be to reverse the current policy of compromise in the goals of the industry road map and put the program back on the original roadmap.

Figure 3. Plasma Etching on the DEB-co-TEB 450nm thin films. a) contact UV lithography technique to write the etching pattern, b)after etch by RIE O_2 plasma, c) remove photoresis

The unusually high modulus and hardness values observed suggest that they arise from the formation of a layered aromatic structure, which implies an ability to match the CTE of substrates such as Si and Ta. This would sharply reduce stresses at the interfaces. However in the perpendicular direction one would have to design around a somewhat increased CTE. Efforts to further enhance the adhesive characteristics of the dielectric to Si and Ta appear promising incorporating thermally resistant polyester groups into the diacetylene oligomer. [5] The potential exists to further improve the thermal conductivity over the measured value of 0.33 W/m K. The very high modulus and hardness values already observed suggest that the formation of the layered aromatic structure may tend to match the CTE of substrates such as Si and Ta minimizing stresses at the interface. These include the coefficient of thermal expansion (CTE), biaxial thermal stress, and thermal conductivity. Thus the CTE in the perpendicular direction is $2.0*10^{-5}K^{-1}$ and in planar direction $8.0*10^{-6}K^{-1}$(Table II). The low CTE provides a better match to the Si substrate which also to minimize interfacial stress and greatly enhance the reliability of the microprocessors.

Table II. Coefficient of Thermal Expansion of cured DEB-co-TEB

Film Thickness 510nm	In-Plane (2"Wafer)	Normal-to-Plane
Intrinsic DEB-co-TEB	$8.0*10^{-6}$ K^{-1}	$2.0*10^{-5}$ K^{-1}
DEB-co-TEB With 20% Porogen	$7.3*10^{-6}$ K^{-1}	$1.7*10^{-5}$ K^{-1}

Porogen: Polystyrene

CONCLUSIONS

The research at its present stage of development appears to offer the avenue for designing an ultra low k dielectric, which can display the desired modulus and hardness values to withstand the process of CMP. Initial experiments with oxygen etching suggest a high probability of success for achieving vertical profiles. This system can potentially be implemented directly into an assembly line for spin coating of dielectric followed by porogen evolution to yield the desired dielectric. The forrmation of the layered aromatic structure may tend to match the CTE of substrates such as Si and Ta minimizing stresses at the interface. The near and long term significance to the semiconductor industry would be to reverse the current policy of compromise in the goals of the industry road map and put the program back on the original roadmap.

REFERENCES

1. ITRS Roadmap for Semiconductors Interconnect (2005)
2. R.J.O.M. Hoofman, G.J.A.M. Verheijden, J. Michelon, F. Iacopi,Y. Travaly, M.R. Baklanov, Zs. T kei, G.P. Beyer Microelectronic Engineering (2005), 80, 337–344
3. C.M. Garner G. Kloster, G. Atwood, L. Mosley, A.C. Palanduz, Microelectronics Reliability (2005), 45, 919–924
4. Yongqing Huang, John McCormick, James Economy. Polymers for Advanced Technologies (2005), 16(1), 1-5
5. Yongqing Huang, James Economy. Macromolecules (2006), 39, 5, 1850-1853.
6. Ken Schroeder. Future FAB international (2005), 19,18-21
7. Thomas Neenan, Matthew Callstrom, Louis Scarmoutzos, George Whitesides. Macromolecules (1988), 21, 3528-3530
8. James Economy. High Temperature Polymers for Electronic Applications, Plenum Press, Contemporary Topics in Polymer Science (1984), 5, 351-375
9. Yongqing Huang. PhD Dissertation: Design of Novel High Temperature Thermosetting Resins Tailored to Specific Needs (2004)
10. A.C. Diebold. Handbook of Silicon Semiconductor Metrology (2001)
11. URL: http://www.physorg.com/news5067.html

New Materials and Characterization II

Mater. Res. Soc. Symp. Proc. Vol. 1134 © 2009 Materials Research Society 1134-BB03-04

Polyaniline Nanostructures for Hydrogen Storage Applications

Michael U. Niemann[†], Sesha S. Srinivasan[†], Ayala R. Phani[‡], Ashok Kumar[†], D. Yogi Goswami[†] and Elias K. Stefanakos[†]

[†]Clean Energy Research Center, College of Engineering, University of South Florida, Tampa, FL 33620, USA
[‡]Nano-RAM Technologies, 98/2A Anjanadri, 3rd Main, Vijayanagar, Bangalore 5600040, Karnataka, India

I. ABSTRACT

Polyaniline nanostructures with fibrous and granular morphology have been synthesized using a chemical process in presence of surfactants as dopants. The reversible hydrogen sorption characteristics, namely sorption kinetics measurements, were performed on these polyaniline (PANI) nanostructures at high hydrogen pressures and room temperatures. The rapid uptake and release of hydrogen up to >4 wt.% was observed in these nanostructures, which may be due to their unique morphologies as explored from microstructural observations.

II. INTRODUCTION

The conducting polymer polyaniline is considered to have potential promise for various applications because of its extraordinary properties of electrical and optical behavior. Polyaniline is a well studied polymer and its synthesis varies greatly [1-3], producing various structures. It was recently reported that polyaniline could store as much as 6 to 8 weight percent (wt.%) of hydrogen [4]; however, the reproducibility was not established [5]. Though many controversial results were reported in terms of hydrogen uptake [6-8] in polymer nanocomposites, there are still a number of parameters, tailor-made properties, surface morphologies and their correlation with hydrogen sorption behavior to be investigated before these materials can be commercially deployed for on-board hydrogen storage. Similarly, nanotubes [9, 10] or nanofibers [11] have attracted more interest because of their novel properties and wide potential application for nanometer-scale engineering applications. It is known that nanofibrillar or nanospherical morphologies significantly improve the performance of polyaniline in many conventional applications involving polymer interactions with its environment [12]. This leads to faster and more responsive chemical sensors [13], new organic/polyaniline nanocomposites [14] and ultra-fast non-volatile memory devices [15].

In the present study, we report the synthesis of polyaniline nanofibers and nanospheres by chemical process. While there has been extensive research into synthesizing nanofibers and nanospheres [16-17], the synthesis techniques have been altered slightly in the present study. These as-synthesized nanostructures were characterized using Fourier Transform Infrared (FTIR) spectroscopy. Hydrogen adsorption and desorption behavior in these nanostructures were estimated by high pressure hydrogen sorption measurements. Microstructural changes due to hydrogen sorption were observed by Scanning Electron Microscopy (SEM).

III. EXPERIMENTAL DETAILS

We have synthesized the polyaniline nanofibers (PANI-NF) and PANI with granular structure, referred to as nanospheres (PANI-NS), by oxidative polymerization of aniline monomer at 0°C in an ice bath using ammonium persulfate as the oxidant in the presence of surfactant. Aniline, ammonium persulfate, dodecyl benzene sulfonic acid, acrylmethylpropyl sulfonic acid, and camphorosulfonic acid are used as received from Sigma-Aldrich. Sulfonic acid based surfactants as the dopant and ammonium persulfate as the oxidant were used in the present synthesis of polyaniline nanofibers (see Figure 1). Polyaniline nanofibers were synthesized by oxidative polymerization of aniline monomer at 0°C in an ice bath using ammonium persulfate as the oxidant in the presence of surfactant. Calculated quantities of aniline monomer (0.005 mol) were mixed with 10 ml of distilled water and stirred using a magnetic stirrer for 10 minutes. Meanwhile, calculated quantities of camphorosulfonic acid (0.0025 mol) and ammonium persulfate (0.005 mol) were dissolved separately in 5ml distilled water and stirred for 10 minutes in an ice bath. The surfactant solution was first added drop wise into the aniline monomer aqueous solution followed by the previously cooled oxidant solution, also drop wise. The mixture was then allowed to react for 15 hours in an ice bath, while being stirred constantly. The precipitate was filtered and washed several times with distilled water until the filtrate solution was clear, followed by a methanol wash to terminate the polymerization reaction. Finally, the PANI was dried in vacuum at room temperature for 24 hours. Later, the vacuum-dried precipitate was annealed at 125°C for 3 hours. The dried polyaniline was characterized and tested for the hydrogen uptake and release measurements.

Similarly, PANI nanospheres were synthesized with only few modifications. Instead of dissolving the aniline and ammonium persulfate in water, however, they were dissolved in a 1M solution of HCl at 0°C (also 0.005 mol of aniline and ammonium persulfate). The ammonium persulfate solution was then added drop wise to the aniline solution and allowed to mix, while stirring, for 5 hours at 0°C in an ice bath. The camphorosulfonic acid solution, as described for the synthesis of the nanofibers, was then added drop wise to the solution and stirred for 15 hours at 20°C. Finally, the PANI was again washed with DI water, followed by a methanol wash as previously described. The sample was again dried in vacuum at room temperature and then annealed at 100°C for 1 hour.

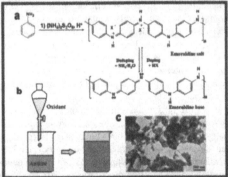

Figure 1. Schematic synthesis set up for the polyaniline nanofibers and nanospheres

The carbon-hydrogen (C-H) and carbon-nitrogen bond stretches of the PANI-NF and PANI-NS were compared via Perkin-Elmer Spectrum One FTIR spectrometer. This instrument operates in a single-beam mode and is capable of data collection over a wave number range of 370-7800 cm^{-1} with a resolution of 0.5 cm^{-1}. The PANI-nanostructured samples were pelletized and sealed in a specially designed KBr cell for infrared measurements.

The microstructure of the as-synthesized PANI nanostructures was studied by Hitachi S800 scanning electron microscope (SEM). A fixed working distance of 5mm and a voltage of 25 kV were used. Sample preparation for the SEM measurement was carried out inside the glove box by covering the sample holder with parafilm for minimal exposure to oxygen while transferring it to the secondary emission chamber. EDAX Genesis software was used to analyze the SEM images.

The volumetric hydrogen sorption measurements of PANI-NF and PANI-NS were measured using Sievert's type apparatus. Room temperature hydrogen absorption was executed at a high pressure (H_2 ~80 bar) with a pre-calibrated reservoir. These isothermal volumetric measurements were carried out by Hy-Energy's PCTPro 2000 sorption equipment. This fully automated Sievert's type instrument uses an internal PID controlled pressure regulator with maximum pressure of 170 bar. It also includes five built-in and calibrated reservoir volumes of 4.66, 11.61, 160.11, 1021.30 and 1169.80 ml. The volume calibration with and without the sample was performed at a constant temperature with an accuracy of ±1 °C using helium. The software subroutines for hydrogen purging cycles, leak test, kinetics, PCT and cycling were performed by the HyDataV2.1 Lab-View program. The data collected for each run were analyzed using the Igor Pro 5.03 program with a built-in HyAnalysis Macro.

IV. RESULTS AND DISCUSSION

Fourier Transform Infrared (FTIR) spectra of PANI-NF and PANI-NS prepared from chemical method are shown in Figure 2. The major bonding environment remains unchanged for both nanofibrous and nanospheres polyaniline structures. The presence of two bands in the vicinity of 1500 cm^{-1} and 1600 cm^{-1} are assigned to the non-symmetric C6 ring stretching modes. The higher frequency vibration at 1600 cm^{-1} is for the quinoid rings, while the lower frequency mode at 1500 cm^{-1} depicts the presence of benzenoid ring units. Furthermore, the peaks at 1250 cm^{-1} and at 800 cm^{-1} are assigned to vibrations associated with the C-N stretching vibration of aromatic amine out of plane deformation of C-H of 1,4 disubstituted rings. The aromatic C-H bending in the plane (1167 cm^{-1}) and out of plane (831 cm^{-1}) for a 1,4 disubstituted aromatic ring indicates a linear structure.

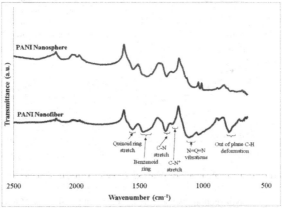

Figure 2. FTIR spectra of PANI-NF and PANI-NS obtained from chemical process

Figure 3 represents the scanning electron micrographs of polyaniline nanofibers and nanospheres obtained from the chemical process. It is evident that the slight modifications in synthesis conditions produce either nanofibrous or nanospherical structures. The nanospheres are uniformly distributed throughout the surface of the PANI, whereas the nanofibers exhibit growth from the base structure and are not dense throughout the surface. While the yield of these synthesis parameters is not 100% nanofibers or nanospheres, there is nevertheless a significant difference in the two microstructures. There are, in fact, regions of the sample that exhibit a higher amount of nanofibers or nanospheres, though, the more typical SEM images are shown in this manuscript.

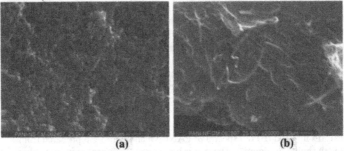

(a)   (b)

Figure 3. SEM micrographs of polyaniline (a) nanospheres and (b) nanofibers before hydrogen sorption

Figure 4 demonstrates the initial hydrogen uptake of PANI-NF and PANI-NS with respect to time. From this figure it is seen that rapid absorption was achieved (i.e. 95% of total capacity in less than 2 hours) by these nanostructures.

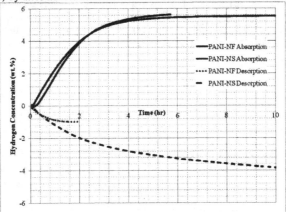

Figure 4. Hydrogen sorption kinetics in polyaniline nanostructures at 30°C

Almost 4-5 wt.% of hydrogen uptake was observed during first cycle of adsorption at 30°C. However, during hydrogen desorption, the nanospheres show a higher amount of hydrogen being released, when compared to the PANI nanofibers. This may be due to the more uniform particle size distribution in PANI nanospheres as compared to the nanofibrous morphology (see Figure 3). The rate at which hydrogen is released, however, is significantly slower than the hydrogen absorption, while the capacity is also lower, indicating that some of the hydrogen that is absorbed is not released during the desorption measurement. However, one has to consider that the desorption is performed at a relatively low temperature, and one can expect a higher amount of hydrogen released if the sample were heated. This is currently under investigation and the results will be forthcoming.

After several hydrogen sorption cycles, these samples exhibit agglomeration and coagulation due to hydrogen interaction on the surface which is shown in Figure 5. Neither the nanospherical nor nanofibrous morphology is seen from these microstructures which ensure effective hydrogen reaction which produces micro-cracks on the surface. Figure 6 shows the FTIR spectrum of the nanofibers before and after hydrogen cycling. It is clear that there is no chemical change in the PANI (e.g. from emeraldine base to leucoemeraldine due to hydrogen reduction). Currently, we are executing long term cycling studies in these polyaniline nanostructures and the results will be forthcoming.

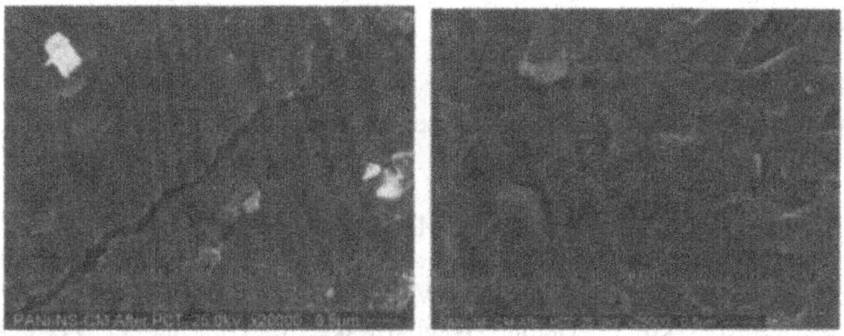

               (a)   **(b)**

Figure 5. SEM micrographs of PANI (a) nanospheres (b) nanofibers after hydrogen sorption

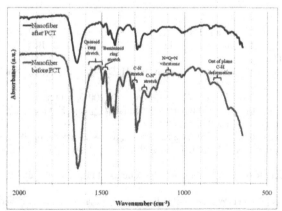

Figure 6. FTIR spectra of PANI nanofibers before and after hydrogen cycling

V. CONCLUSIONS

In summary, we have investigated polyaniline nanofibers (PANI-NF) and nanospheres (PANI-NS) synthesized by chemical templating technique using various surfactants as dopants and ammonium persulfate as the oxidant. The characteristics of PANI-NF are understood regarding their structure, microstructure, and bonding behaviors. The rate of hydrogen sorption at room temperature during the initial cycle shows rapid (95% hydrogen storage capacity (~4-5 wt.%) absorbed in less than 2 hours). Moreover, the PANI nanospheres demonstrate excellent reversibility in the desorption cycle when compared to PAni nanofibers. This behavior may be due to the uniform particle distribution with high surface area of nanospheres for hydrogen reaction. However, after several hydrogen sorption cycles, the morphological changes due to hydrogen interaction on the surface created agglomeration and micro-crack formation which needs further understanding of the reaction mechanism.

VI. ACKNOWLEDGMENTS

Authors wish to acknowledge the US Department of Energy (Hydrogen Fuel Initiative code: **DE-FG36-04G014224**) for funding. USF NanoTech Research Facility for analytical studies is gratefully acknowledged.

VII. REFERENCES

[1] N. Chandrakanthi and M.A. Careem: Polymer Bulletin Vol. 45 (2000), p.113

[2] G. Li, Z. Wang, G. Xie, H. Peng, and Z. Zhang: Journal of Dispersion Science and Technology Vol. 27 (2006), p. 991

[3]S. Xing, H. Zheng, and G. Zhao: Synthetic Metals Vol. 158 (2008), p. 59

[4] S.J. Cho, K.S. Song, J.W. Kim, T.H. Kim and K. Choo: Preprints of Symposia - American Chemical Society, Division of Fuel Chemistry Vol. 47 (2002), p. 790

[5] B. Panella, L. Kossykh, U. Dettlaff-Weglikowska, M. Hirscher, G. Zerbi and S. Roth: Synthetic Metals Vol. 151 (2005), p. 208

[6] N.B. McKeown, P.M. Budd and D. Book: Macromol. Rapid Commun. Vol. 28 (2007), p. 995

[7] S.J. Cho, K. Choo, D.P. Kim and J.W. Kim: Catalysis Today Vol. 120 (2007), p. 336

[8] M.U. Jurczyk, A. Kumar, S. Srinivasan and E. Stefanakos: Int. J. Hydrogen Energy Vol. 32 (2007), p. 1010

[9] J. Germain, J.M.J. Frechet and F.J. Svec: Mater. Chem. Vol. 17 (2007), p. 4989

[10] A.C. Dillon, K.M. Jones, T.A. Bekkedahl, C.H. Kiang, D.S. Bethune and M.J. Heben: Nature Vol. 386 (1997), p. 377

[11] A. Nikitin, X. Li, Z. Zhang, H. Ogasawara, H. Dai and A. Nilsson: Nano Lettters Vol. 8 (2008), p. 162

[12] Y. Wang and X.J. Jing: J. Phys. Chem. B Vol. 112 (2008), p.1157

[13] A.Z. Sadek, A. Trinchi, W. Wlodarski, K. Kalantar-zadeh, K. Galatsis, C. Baker and R.B. Kaner: IEEE Sensors Vol. 3 (2005), p. 207

[14] A.A. Athawale and S.V.J.J. Bhagwat: J. Appl. Polymer Science Vol. 89 (2003), p. 2412

[15] Y. Yang, J. Ouyang, L. Ma, R.J. Tseng and C.-W. Chu: Adv. Funct. Mater. Vol. 16 (2006), p. 1001

[16] J. Huang and R.B. Kaner: Angewandte Chemie International Edition, Vol. 43 (2004), p.5817

[17] J. Huang, S. Virji, B.H. Weiller, and R.B. Kaner: Journal of American Chemical Society, Vol. 125 (2003), p.314

Mater. Res. Soc. Symp. Proc. Vol. 1134 © 2009 Materials Research Society   1134-BB03-06

Operation Characteristics of Ionic Polymer-Metal Composite Using Ionic Liquids

Kunitomo Kikuchi[1] , Masafumi Miwa[2] and Shigeki Tsuchitani[3]
[1]Graduate School of Systems Engineering, Wakayama University, 930, Sakaedani, Wakayama, 640-8510, JAPAN
[2] Institute of Technology and Science, The University of Tokushima, 2-1, Minamijosanjima-cho, Tokushima, 770-8506, JAPAN
[3]Department of Opto-Mechatronics, Faculty of Systems Engineering, Wakayama University, 930, Sakaedani, Wakayama, 640-8510, JAPAN

ABSTRACT

Operating characteristics of ionic polymer-metal composite (IPMC) using ionic liquids (ILs) as solvent were measured in air. Among three kinds of IL (1-ethyl-3-methyl-imidazolium tetrafluoroborate (EMIBF$_4$), 1-buthyl-3-methyl-imidazolium tetrafluoroborate (BMIBF$_4$), and 1-buthyl-3-methyl-imidazolium hexafluorophosphate (BMIPF$_6$)), IPMC using BMIPF$_6$ exhibited the largest curvature of 3.3 m^{-1} under an application of a rectangular wave voltage (±2.0V) with a frequency of 0.5Hz. It was found that the behavior of IPMC using ILs changed depending on environmental humidity. The maximum curvature during the voltage application, and the initial response speed of curvature were larger with increasing environmental humidity and voltage height. Time dependence of the curvature well corresponds to that of the transported charge due to movement of counter ions. Therefore, it is possible to explain the curvature response by the behavior of the transported charge.

INTRODUCTION

Recently, polymer actuators are attracting much attention due to their special characteristics. Ionic polymer-metal composite (IPMC) is one of such polymer actuators which has flexibility, light weight, and low driving voltages lower than 2 or 3 V, and has a potential application as soft robotic actuators [1]. It consists of a polyelectrolyte membrane and thin noble metal electrodes formed on the both surfaces of the membrane. By applying voltages between the both electrodes, IPMC bends toward higher potential side. Conventional IPMC cannot work in dry air environment because of decrease of ionic conductivity of the polyelectrolyte membrane in IPMC. Recently, it was reported that IPMC using ionic liquid (IL) as a solvent could work in dry air environment [2-4].

In this study, we fabricated Nafion®-based IPMC using IL, and drive voltage frequency response and step voltage response of bending curvature, and complex impedance were evaluated at controlled environmental humidity.

EXPERIMENTAL DETAILS

Preparation of IPMC using Ionic liquid

Sample IPMC was prepared by connecting electrodes by a chemical plating method [5]. This plating consists of the next two processes.

(1) Adsorption process: A polyelectrolyte membrane was immersed in an aqueous solution of a gold complex to exchange ions for 24 hours at room temperature.

(2) Reducing process: The adsorbed cationic gold complex in the membrane was reduced with an aqueous solution containing a reducing agent for 5 hours at 60 C.

In this study, Nafion® 117 (Du Pont, PFSA membrane), [Au(phen)Cl$_2$]Cl (phen = 1, 10 - phenanthoroline), and Na$_2$SO$_3$ were used as the polyelectrolyte membrane, the cationic gold complex, and the reducing agent, respectively.

After plating, IPMC was cut into rectangle pieces (2 × 14mm). Each peace was immersed in the mixture of deionized water and 0.2M ionic liquid to exchange counter ions at room temperature. 1-ethyl-3-methyl-imidazolium tetrafluoroborate (EMIBF$_4$), 1-buthyl-3-methyl-imidazolium tetrafluoroborate (BMIBF$_4$), and 1-buthyl-3-methyl-imidazolium hexafluorophosphate (BMIPF$_6$) were used as ionic liquid. EMIBF$_4$ and BMIBF$_4$ are water-miscible. BMIPF$_6$ is water-insoluble. After counter ions were exchanged, IPMC was dried in a vacuum desiccator. Then, IPMC was dried for 60 minutes at 90 C before measurement.

Measuring method of bending characteristics of IPMC

Bending curvature of IPMC was measured as follows.

(a) The position of 2mm from the end of the dried IPMC was hold between two gold electrodes of a connecting clamper in a chamber in which humidity was controlled by using a humidity generator and temperature was 25C.

(b) After humidity in the chamber was stabilized, IPMC was kept in this environment for 30 minutes.

(c) The displacement of IPMC was measured by using a laser displacement sensor (Keyence, LK-G85), when rectangle wave voltages or step voltage were applied. The measured position was 2mm from the other end of the IPMC.

(d) From the measured displacement (Δ), the bending curvature $\left(\dfrac{1}{R} - \dfrac{1}{R_0}\right)$ was calculated by the following eq. (1).

$$\frac{1}{R} - \frac{1}{R_0} = \frac{2\Delta}{l^2 + \Delta^2} \quad (\Delta << l) \tag{1}$$

where l=10mm is the length from the measurement point to the clamp point, R_0 and R are the radius of curvature of IPMC before and during measurement, respectively.

To evaluate frequency response of curvature, frequency dependence of curvature of IPMC under application of rectangle wave voltage, were measured in a frequency range from 0.05 to 5.0Hz. In this experiment, the amplitude of the applied voltage was 2.0 V and the measurement humidity was 50%RH.

In addition, to evaluate effects of environmental humidity on step voltage response of curvature, a step voltage with a height of 0.5V was applied to IPMC in environmental humidity of 20, 40, 60, and 80% RH for 300s and time variation of displacement was measured. Furthermore, to evaluate effects of voltage height on step voltage response of curvature, step voltages, with height of 0.5, 1.0, 1.5, and 2.0V were applied to IPMC in an environment with humidity of 60%RH for 300s, and time variation of displacement was measured.

Measuring method of complex impedance of IPMC

To evaluate the electrical characteristics of IPMC, frequency dependence of complex impedance of IPMC was measured by the next way.
(a) The dried IPMC was hold between titanium electrode of a connecting clamper in a chamber (temperature: 25C). The bending of IPMC was suppressed by the clamper.
(b) After the humidity in the chamber was stabilized (20, 40, 60, and 80%RH), IPMC was kept in this environment for 30 minutes.
(c) The complex impedance of IPMC was measured by using an impedance analyzer (HEWLETT PACKARD, 4192A). Measurement frequency was changed from 5Hz to 13MHz. Measurement voltage was $1.0V_{rms}$.

RESULTS

Frequency response characteristics of IPMC

Frequency response characteristics of IPMC are show in Figure 1. Curvature of IPMC using IL was larger in order of $BMIPF_6$, $EMIBF_4$ and $BMIBF_4$ in all operating frequency (0.05-5.0 Hz) at 25C and 50%RH. Compared with the previous study concerning Flemion®-based IPMC using $BMIBF_4$[4], the present Nafion®-based IPMC using $BMIBF_4$ exhibited a curvature of 0.8 m^{-1}, which was approximately half of that of the Flemion®-based IPMC ($1.5m^{-1}$) which was evaluated under an application of a rectangular wave voltage (±2.0V) with a frequency of 0.5Hz. On the other hand, the curvature of Nafion®-based IPMC using $BMIPF_6$ was almost same as that of Flemion®-base IPMC using $BMIBF_4$. In the previous study concerning Flemion®-based IPMC, the one using $BMIBF_4$ had larger displacement than that using $BMIPF_6$. This difference of the performance between Nafion®-based and Flemion®-based IPMCs using the same kinds of ILs might be based on difference in molecular structure and electrical affinity to ILs of the both ion exchange materials.

Humidity dependence of bending characteristics of IPMC

In the case IPMC using $EMIBF_4$, effects of environmental humidity on time dependence of curvature, transported charge and variation in curvature after an application of a step voltage with a height of 0.5V, are shown in Figure 2. Transported charge represents the total charge

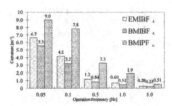

Figure 1. Curvature of IPMC using ionic liquids under application of rectangle wave voltage with an amplitude of 2.0V. It was larger in order of $BMIPF_6$, $EMIBF_4$ and $BMIBF_4$ in all operating frequency (0.05-5.0Hz) at 25C and 50%RH.

transported by cations during actuation, and was calculated by integrating the current across the thickness of IPMC. The variation in curvature was derived from the slope of the time dependence curve of curvature. Moreover, in the case IPMC using EMIBF₄, effects of voltage height on time dependence of curvature, transported charge, and variation in curvature after applications of various height (0.5-2.0V) of step voltages at a constant humidity (60%RH) are shown in Figure 3.

The maximum curvature during the voltage application for 300s was larger with increasing environmental humidity and voltage height as shown in Figs. 2-(a) and 3-(a). In addition, slope of response curves at initial stage (variation in curvature) increased with increase in the humidity and voltage height as shown in Figs. 2-(b) and 3-(b).

Complex impedance of IPMC

Figure 4 shows Cole-Cole plots of IPMCs using the three kinds of ILs in a frequency range over 5Hz. Horizontal axis is resistance $Re(Z)$, i.e., real part of impedance Z, and vertical axis is reactance $Im(Z)$, i.e., imaginary part of impedance. The complex impedances were plotted in semicircular shape, in all measurement conditions. This result indicates that the equivalent circuit of IPMC is expressed by a parallel combination of resistance R_p and capacitance C_p in this

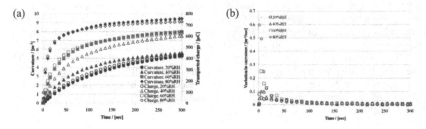

Figure 2. Effects of environmental humidity on time dependence of (a) curvature and transported charge, and (b) variation in curvature of IPMC using EMIBF₄ when a step voltage with a height of 0.5V was applied

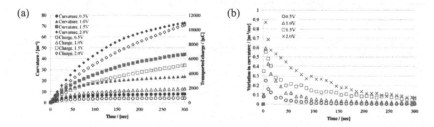

Figure 3. Effects of voltage height on time dependence of (a) curvature and transported charge, and (b) variation in curvature of IPMC using EMIBF₄, when step voltages with various heights were applied at 60%RH

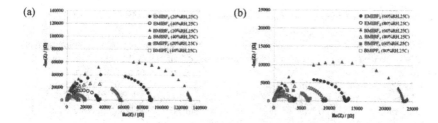

Figure 4. Cole-Cole plots of IPMC at (a) 20 and 40%RH, and (b) 60 and 80%RH

frequency range. The value of R_p of IPMC, which is derived from the diameter of the semicircle, decreased with increase in environmental humidity. This means that ionic conductivity of the Nafion® membrane increased with increasing environmental humidity. In the samples using the three kinds of ILs, the sample using BMIPF₆ had the smallest resistance element.

DISCUSSION

In Figure 2-(a) and 3-(a), time variation of the curvature well corresponds to that of the transported charge. Therefore, it is possible to explain the curvature response by the behavior of the transported charge. In IPMC, charge is transported thorough ionic polymer (Nafion®). The behavior of the electrode-ionic polymer interface is analogous to that of a capacitor. Therefore, we regard IPMC using ILs as a serial combination of a resistor R_s and capacitor C_s as a first approximation, in low frequency range (<~5Hz). Where, R_s and C_s are charge transfer resistance, which relates to ionic conductivity, and electrode-ionic polymer interface capacitance, respectively. On the other hand, in the frequency range higher than ~5Hz, IPMC was expressed by the parallel combination of R_s and C_s, as described in the previous section.

Comparing the results of complex impedance measurements (Figure 4) with those of the variation in curvature (Figure 2-(b)), the initial response speed (variation in curvature just after voltage application) was larger in IPMCs having lower R_p and at higher humidity at which R_p was lower. In the previous study [6], it was reported that ionic conductivity of Aciplex® (Asahi Kasei Chemicals, PFSA membrane), which is a polyelectrolyte as Nafion®, increased as the humidity increased. Therefore, R_s is regarded as the charge transfer resistance, i.e., $R_s = R_p$. So, it is said implies that the initial curvature response of IPMC is mainly determined by the value of the charge transfer resistance, because the charge transfer resistance determines the charge accumulation speed to the electrode-ionic polymer interface capacitor C_s in the initial stage. The same phenomena were confirmed, when the applied voltage was changed as shown in Fig 3. Because counter ions move by electrostatic force, the moving speed of counter ions increases with increasing applied voltage and the charge accumulation speed would increase.

As for the maximum curvature at 300s after the voltage application, it was larger with increasing environmental humidity and the applied voltage. Since we assumed the serial combination of R_s and C_s as the equivalent circuit of IPMC at the lower frequency range, the maximum curvature is determined the charge amount accumulated to C_s. Therefore, it is said that C_s increases with increase in environmental humidity.

In the previous study [1, 5, 7-9], it is reported that Nafion®-based IPMC containing metallic cations bended rapidly reaching the maximum displacement to the anode side, then the curvature reversed gradually. This bending back process was explained by water permeation due to the pressure gradient created in the bent IPMC [8]. However, the present IPMCs using ionic liquids, IPMC bent monotonously without the backward relaxation. In general, ionic radius of cations of ionic liquid is almost the same or larger size than that of hydrated metallic cations [1, 11]. Therefore, it is thought that cations of ionic liquids had pumping effects to water molecules as reported in the previous study [1], because size of ion channel of IPMC is almost equal to or larger than the ion radius of ionic liquid.

CONCLUSIONS

The operating characteristics of IPMC using ionic liquids (EMIBF$_4$, BMIBF$_4$, and BMI-PF$_6$) as solvent were measured in air. The sample using BMIPF$_6$ exhibited the largest displacement among different sample, by application of rectangle wave voltage of ±2.0V at operating frequency over 0.05 Hz. The maximum curvature and the initial response speed of curvature were larger with increasing environmental humidity and applied voltage. In the present IPMC using ionic liquid, it was found that displacement response was associated with the transported ionic charge in IPMC. The experimental results imply that moving speed and the amount of transported charge of counter ions in IPMC increased with increasing humidity and applied voltage.

ACKNOWLEDGMENTS

This work was supported in part by KAKENHI (No. 18560249) from Japan Society for the Promotion of Science (JSPS). The authors would like to express his thanks to Dr. Kinji Asaka of National Institute of Advance Industrial Science and Technology (AIST) for his technical support for fabricating IPMC.

REFERENCES

1. K. Asaka, in *Soft Actuator kaihatsu no saizensen*, edited by Y. Osada, (NTS, Tokyo, 2004) pp. 76-95; K. Onishi, ibid, pp. 96-118(in Japanese).
2. J. Wang, C. Xu, M. Taya, and Y. Kuga, *Proceedings of SPIE*, **6168**, 61680R (2006).
3. K. Asaka, K. Mukai, Y. Ogawa, K. Kiyohara, T. Fukushima, A. Kosaka and T. Aida, *Proceedings of SI2007*, pp.215-216 (2007).
4. M. Bennett and L. Donald., *Proceedings of SPIE*, **5759**, pp. 506-517 (2005).
5. N. Fujiwara, K. Asaka, Y. Nishimura, K. Oguro, and E. Torikai, *Chem. Mater.*, **12**, 6, pp.1750- 1754 (2000).
6. Y. Hashimoto, N. Sakamoto, and H. Iijima, *Koubunshi Ronbunshu*, **63**, 3, pp. 166-173 (2006).
7. K. Asaka, K. Oguro, Y. Nishimura, M. Mizuhata, and H. Takenaka, *Polym. J.*, **27**, 4, pp. 436-440(1995)
8. K. Asaka, K. Oguro, *Journal of Electroanalytical Chemistry*, 480, pp.186-198 (2000).
9. S. Nemat-Nasser and Y. Wu, *J. Appl. Phys*, **93**, 9, pp.5225-5267 (2003).
10. T. Yamaue, H. Mukai, K. Asaka, and M. Poi, *Macromolecules*, **38**, pp.1349-1356 (2005).
11. R. Kawano, H. Tokuda, T. Katakabe, H. Nakamoto, H. Kokubo, S. Imabayashi, and M. Watanabe, *Koubunshi Ronbunshu*, **63**, 1, pp. 31-40 (2006).

Mater. Res. Soc. Symp. Proc. Vol. 1134 © 2009 Materials Research Society   1134-BB03-10

Dielectric Breakdown of Transformer Insulation Materials Under Cryogenic and Room Temperatures

H. Rodrigo[1], W. Baumgartinger[1], A. Ingrole[2], Z.(Richard) Liang[2], D.G. Crook[1] and S.L. Ranner[1]

[1]Center for Advanced Power Systems, Florida State University, 2000 Levy Avenue, Tallahassee, FL 32310

[2]Department of Industrial and Manufacturing Engineering, FAMU-FSU College of Engineering, 2525 Pottsdamer Street, Tallahassee, FL 32310

ABSTRACT

The results of electrical breakdown field measurements on Thermavolt[R] , a material used widely in the manufacture of power transformers is presented. The performance of this material is compared with others that consist of polymer resins that have been modified by the addition of nano-particles. The polymer used is Polymethylmethacrylate (PMMA) with the addition of Barium Titante (BTA). The breakdown voltage measurements have been conducted under 60 Hz AC high voltage and 1.2/50 µs lightning impulse high voltage. The electrode configuration gives a uniform field with 25 mm diameter Stainless Steel electrodes with Bruce profile. Measurements presented are at room temperature (293 K) and at liquid Nitrogen temperature (77 K). The results of AC and impulse breakdown show a marked improvement in the performance of Thermavolt as the temperature is reduced from 293 K to 77 K. However, the differences in breakdown values under AC and impulse voltages for the nanocomposite PMMA + BTA was marginal. Neat PMMA showed a very slight improvement in the breakdown value under AC voltage from 293 K to 77 K. The breakdown voltage under impulse conditions showed a very significant decrease for neat PMMA as the temperature was reduced to 77 K. All three materials studied exhibited higher dielectric losses at 77 K than at 293 K. SEM images of the samples show the shock propagation patterns at the two different temperatures for PMMA + BTA nanocomposite under the two waveform conditions.

1.0 INTRODUCTION

Dielectrics form an important part in the design and manufacture of transformers, electrical machines, coils and cables. The electrical insulation of transformers pose many challenges, they become more complex at cryogenic temperatures [1,2,3]. In a transformer there is insulation between the turns of a coil, the core and the coils and between the coils. The turn to turn insulation has to be pliable and flexible. For the other areas the insulation has to be more rigid and in cases of those operating at cryogenic temperatures is called upon to carry some mechanical load. The turn to turn insulation does not encounter very high voltages, generally the voltages are of the order of a few kV and at most a few 10s of kV. However, as the insulation is wrapped around the conductors, and often there could be more than one layer of insulation around each conductor, butt gaps are formed. These butt gaps are points of weakness in dielectric strength, this is particularly the case with HTS transformers cooled with liquid Nitrogen operating between 30 K and 50 K under super cooled conditions [2]. In the present work we have studied the dielectric properties of three materials, namely; Thermavolt, Polymethylmethacrylate (PMMA) and PMMA + 5wt% of Barium Titanate nano-particles (PMMA + BTA). Thermavolt is a material that is in wide usage for turn to turn coil insulation in the manufacture of power transformers. PMMA is widely used as a dielectric material at room temperature. Composite PMMA/BTA was made to study the effect of BTA nano-particles on the dielectric properties of neat PMMA.

2.0 EXPERIMENTAL ARRANGEMENT AND PROCEDURE
2.1 Sample Preparation
Neat PMMA (Polymethyl Methacrylate) was weighed and placed on a hot die press, the temperature was raised gradually until the temperature reached 230^0C. Pressure was then applied on the press (4.5 Ton). The pressure was maintained and the sample was allowed to cool for between 7 and 10 hours. The clear sheet of PMMA whose thickness was approximately 1mm was then removed from the press and cut using a laser cutter to the required size.

PMMA + BTA (Barium Titanate): 14.25 g of PMMA crystals was dissolved in 150 ml of Chloroform. 750 mg of BTA was mixed with 80 ml of Chloroform on sonicator for 15 mins. The two solutions were mixed and left on sonicator for a further 15 mins. The resulting mixture was allowed to dry for between 24 and 48 hours and placed in a vacuum oven for 12 hours. The semi-solid was then placed on a hot press and the temperature gradually raised until it reached 260^0C, a pressure of 4.5 Tons was then applied to the sample. The press was allowed to cool for between 7 and 10 hours. The sheet of PMMA + BTA was then removed and cut to required size using a laser cutter. The SEM samples were prepared by facturing the samples in liquid nitrogen. A thin Gold coating was applied to the samples by sputtering in a bell chamber. SEM experiments were conducted using a JEOL (7401F) field emission SEM machine.

2.2 AC breakdown measurements
Samples were held between two uniform field electrodes with Bruce profile, made of stainless steel and connected to a high voltage transformer whose rating is 150 KVA.The electrodes were each of diameter 25 mm. The voltage was ramped at a rate of 500 V/Sec until there was a breakdown of the sample. The voltage and waveform at the point of collapse were recorded. For each material ten such measurements were taken at each of the temperatures, 293 K and 77 K. All Thermavolt samples were 0.25 mm in thickness and the other samples studied in this work were of thicknesses ranging from 0.75 mm to 1.00 mm.

2.3 Impulse breakdown measurements.
The samples were held in the same manner as described above. The output of a single stage Haefley impulse generator, rated at 140 kV, 250 J, was applied to the experimental gap in steps of about 5 kV or less until a breakdown was recorded. As before ten such measurements were made for each material at each of the temperatures.

2.4 Tan δ Measurements
A Haefely 2820 Automated Measuring Bridge with a 100 pF standard SF_6 capacitor rated at 100 kV was used for these measurements. The electrodes were connected as described above. The voltage was raised in steps until the voltage was about 80% of the breakdown voltage. At each voltage the measurements were read off the tanδ bridge connected to the experimental set up. The bridge gave the value of tanδ which was a measure of the losses in the dielectric under test. In addition it gave the value of the capacitance of the test piece.

3.0 RESULTS and DISCUSSION
3.1 Weibull Plots for Electrical Breakdown Measurements
3.1.1 AC Breakdown
In the analysis of results for electrical breakdown at both temperatures and under AC and Impulse waveform conditions we have employed Weibull Statistical plots [4,5], and the equation governing the statistics is given as :

$$F(E) = 1 - \exp(-(E/E_0)^\beta)$$

Where E is the electric breakdown field, E_0 is the field for which the cumulative probability is 63.2% and β is the shape factor, which is obtained from the plot of $\log_{10}(-\ln(1-F))$ as a function of $\log_{10} E$. The slope of the regression line fitted to the data points is β.

Figure 1: (a) and (b) show Weibull plots for all materials studied for AC breakdown. at 293 K and 77 K Similarly Figure 2 (a) and (b) show Weibull plots for impulse breakdown at both temperatures 293 K and 77 K.

Under AC breakdown conditions there was a significant improvement in the threshold $(E_0)_{63.2\%}$ for Thermavolt of the order of 44%, where as for the other materials studied the differences were marginal. The slope for Thermavolt was also much shaper at 77 K than was the case at 293 K. Whereas for the other two materials studied the slope at 293 K was about twice that at 77 K. The shape factor was a measure of the spread of data in a given sample set.

3.1.2 Impulse Breakdown

The impulse breakdown was conducted using a positive impulse waveform whose shape was 1.2/50µs, which was a standard lightning impulse. The results indicate that the mechanics of breakdown under this regime was complex. In contrast to the situation under AC other than Thermavolt which has a significant improvement in its breakdown value, which was of the order of 24% when the temperature changed from 293 K to 77 K. PMMA in fact has a 24% decrease in its breakdown value as the temperature was lowered to 77 K from 293 K. The change in PMMA + BTA nanocomposite was marginal being < 10%. In order to understand the mechanics of breakdown in these materials we would have to do more work. Several factors come into play such as particle size, dispersion of the particles in the host resin and the concentration of nano-particles in the finished material. Some work that may help with this analysis is given in [6] where the authors had studied a composite made up of Alumina nano-particles in PMMA. Although they had taken steps to treat the nano-particles prior to mixing with PMMA, so as to increase uniformity of dispersion in the host polymer PMMA. It was reported that the interface between the particles and PMMA which was the host resin created voids. This, if it could be confirmed, will point to a possible cause of decrease in breakdown strength of PMMA + BTA nanocomposite in the present work.

The shape factor under impulse voltages was also markedly different at the two different temperatures presented in this work. For Thermavolt the shape factor has increased by a factor of about 4 between 293 K and 77 K, and for the other two materials the decrement from 293 K to 77 K is of the order of 50%. The results indicate that for PMMA and PMMA + BTA the fluctuations in breakdown strength at 77 K had a larger spread than was the case at 293 K. This result was consistent with the observations of sauers et.al [7] who had studied the breakdown behavior of Stycast.

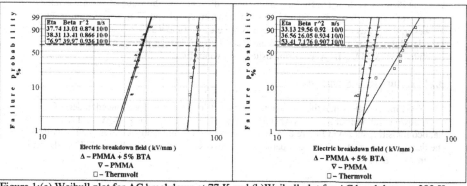

Figure 1:(a) Weibull plot for AC breakdown at 77 K and (b)Weibull plot for AC breakdown at 293 K

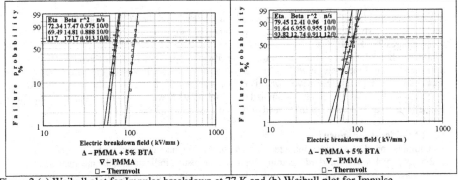

Figure 2:(a) Weibull plot for Impulse breakdown at 77 K and (b) Weibull plot for Impulse breakdown at 293 K

3.13 Tanδ Measurements

Figures 3, 4, and 5 show the variation of tanδ as a function of the electric field (E kV/mm) for Thermavolt, PMMA and PMMA + BTA nanocomposite respectively. In figure 3 (a) we have had to restrict the number of measurements to 3 as the least breakdown value of Thermavolt at 293 K was 6 kV.

We have restricted the maximum of voltage at which tanδ was measured on any sample to between 70% and 80% of its AC breakdown value. In all cases it can be seen that the tanδ value increased as the voltage was increased. This was an indication that there was partial discharge occurring [8] and hence increased the dielectric losses in the dielectric under investigation .Also, at the point of contact close to the edge of the electrode and the dielectric the triple junction effect manifests itself as field enhancement which is a function of the permittivity of the material [9,10,11].

Figure 3 (a) and (b) shows the variation of tan δ as a function of electric field for Thermavolt at 293 K and 77 K respectively. At 10.23 kV/mm the average value of tan δ at 293 K was around 0.035 and the value at 77 K was around 0.011, and as the field was increased to 25.36 kV/mm the values of tan δ were 0.045 at 293 K and 0.022 at 77 K. However, as the field was increased from 25.36 kV/mm to 57.52 kV/mm at 77 K the tan δ value increased rapidly to 0.14, indicating that the dielectric losses also increased.

For neat PMMA as shown in figure 4 (a) & (b) the difference in the value of tan δ when the electric field strength was less than 11 kV/mm between 293 K and 77 K was insignificant. However, there was a significant difference as the electric field was increased, and at the highest field level the losses at 77 K were twice those at 293 K. A similar pattern of behavior was exhibited by PMMA + BTA nanocomposite as shown in figure 5 (a) & (b)

3.14 Fracture surface analysis

The SEM images show that the damage to the dielectric was not as pronounced under AC breakdown conditions at 293 K as when the temperature was 77 K for PMMA + BTA nanocomposites, shown in Figure 6 (a) & (b). However, for breakdown under impulse voltage the damage to the material was comparable at the two temperatures 293 K and 77 K as shown in Figure 6 (c) & (d).

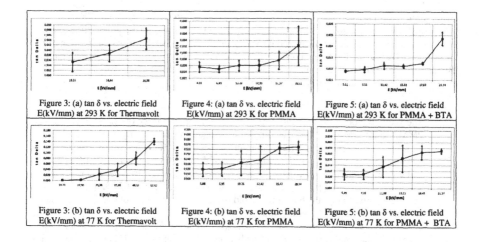

Figure 3: (a) tan δ vs. electric field E(kV/mm) at 293 K for Thermavolt	Figure 4: (a) tan δ vs. electric field E(kV/mm) at 293 K for PMMA	Figure 5: (a) tan δ vs. electric field E(kV/mm) at 293 K for PMMA + BTA
Figure 3: (b) tan δ vs. electric field E(kV/mm) at 77 K for Thermavolt	Figure 4: (b) tan δ vs. electric field E(kV/mm) at 77 K for PMMA	Figure 5: (b) tan δ vs. electric field E(kV/mm) at 77 K for PMMA + BTA

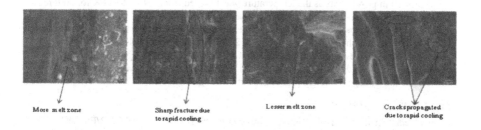

More melt zone   Sharp fracture due to rapid cooling   Lesser melt zone   Cracks propagated due to rapid cooling

Figure 6: (a) PMMA + 5% BTA AC breakdown at 293 K , (b) PMMA + 5% BTA AC breakdown at 77 K, (c) PMMA +5% BTA Impulse breakdown at 293 K, (d) PMMA +5% BTA Impulse breakdown at 77 K.

The breakdown under AC voltage of the dielectric occurs over about two cycles, which amount to approximately 33 ms, in addition there was partial discharge which precedes the breakdown event. In the case at 293K, the experiment was conducted in a bath of transformer oil. The breakdown area shown in the Figure 6 (a), it can be seen that there is a larger melt zone, and as the material was at room temperature the cooling was gradual and there was not a steep temperature gradient, consequently the shock propagation was not as marked. But at 77 K Figure 6 (b), as the material melts because of the voltage, the temperature in that region was very high. As the sample was in cryogenic temperature, the temperature next to the melt down area was very low and it rapidly quenches the material. Because of this rapid cooling there were cracks produced in the breakdown area and sharp fracture edges were formed.

Under impulse voltages the situation was quite different. The total duration from the point of application of the voltage to the point at which breakdown occurs was approximately 2μs to 4μs. Hence the damage to the dielectric was much less pronounced. An additional fact to be taken into account was that there was no partial discharge occurring prior to the breakdown event. Figure 6 (c) & (d) shows the breakdown areas of the material at 293K and 77K respectively. It can be seen that as the application of the voltage was for very short time, there was less amount of melting of the material as compared to AC breakdown. As the cooling of the material at 77K was similar to the previous case, failure occurs due to cracks rather than melting. But at 293 K it can be seen that failure is due to melting of the surrounding material. The melt area is however less than under AC breakdown.

4.0 CONCLUSIONS

1. The Weibull plots show that under AC breakdown conditions the $(E_{b/d})_{63.2\%}$ value for PMMA and PMMA + BTA nanocomposite changes only marginally between 293 K and 77 K. Whereas there was a significant improvement of 44% for Thermavolt as the temperature was lowered to 77 K. The spread of data at 293 K, indicated by the shape factor β, shows that for Thermavolt it was much more than for the other two materials. The lower the value of β the greater the spread of data. However, at 77 K this was completely reversed with about a four fold decrease in the spread of data for Thermavolt and an increase by 50% for the other two materials.
2. Under impulse voltage breakdown the $(E_{b/d})_{63.2\%}$ value for Thermavolt showed a 24.7% increase in the breakdown voltage value between 293 K and 77 K. This was in contrast to the to the decrease in breakdown values for both PMMA and PMMA + BTA nanocomposites. For the PMMA + BTA nanocomposite samples the decrease was marginal. However, for the PMMA samples the decrease was 24.17%. The shape factor showed improvements as the temperature was lowered from 293 K to 77 K for all three materials with PMMA improving by 100%.
3. The dielectric losses given by the tanδ measurements indicate that the losses were greater at 77 K than at the higher temperature of 293 K for all the materials studied.
4. The SEM images lead us to conclude that the damage caused to the materials was greater under AC breakdown voltages than when the breakdown was caused by impulse voltage. Also the damage was more pronounced at 77 K than at 293 K.

ACKNOWLEDGMENTS
Research conducted under DoE contract number D-FC26-07NT43221 and ONR/ESRDC contract number N00014-08-1-0080. Thanks are also due to the High-Performance Material Institute (HPMI) at Florida State University.

REFERENCES
1. J.K.Sykulski, C.Beduz, R.L.Stoll, M.R. Harris, K.F.Goddard and Y.Yang., IEE Proc. Electr. Power appl. 146, 41 (1999)
2. C.S.Weber, C.T.Reis, D.W.Hazelton, S.W.Schwenterly, M.J.Cole, J.A.Demko, E.F.Pleva,S.Metha,T.Golner & N.Aversa.,IEEE Trans Appl.Supercond.,15,2210 (2005)
3. D.R.James, I Saures, A.R.Ellis, E Tuncer, K.Tekletsadik and D.W.Hazelton., IEEE Trans. Appl. Supercond., 17, 1513 (2007)
4. W.Weibull,. Journal of Applied Mechanics, 18, 293 (1951)
5. G.C.Stone, M.Kurtz and R.G.Van Heewijk, IEEE Trans. Electr Insul EI-14, 315 (1979)
6. B.J.Ash, R.W.Siegel and L.S.Schadler, J Poly. Sci.Pt B: Poly. Phy., 42, 4371 (2004)
7. I.Sauers,D.R.James,A.R.Ellis &M.O.Pace, IEEE Trans.Dielec.Electr. Insul, 9, 922 (2002)
8. E.Kuffel,W.S.Zaengl & J.Kuffel, High Voltage Fundamentals, Newnes Press, 412 (2001)
9. T.Takuma and T.Kawamoto, Field , IEEE Trans.Power App.& Syst.PAS103,2486 (1984)
10. H.J.Wintle, IEEE Trans. Electr. Insul, 25,27,(1990)
11. A.Van Roggen, , IEEE Trans Electr. Insul, 25,95,(1990).

Mater. Res. Soc. Symp. Proc. Vol. 1134 © 2009 Materials Research Society   1134-BB03-11

Bilayer microactuator of two kinds of polypyrroles doped with different dopants

Shigeki Tsuchitani, Kosei Chikatani and Kunitomo Kikuchi
Department of Opto-Mechatronics, Wakayama University, 930 Sakaedani, Wakayama-shi,
Wakayama 640-8510, Japan

ABSTRACT

Conjugated polymers such as polypyrrole and polyaniline, have excellent features as actuator, such as flexibility, light weight and large deformation under low application voltage.

We fabricated a bilayer microactuator of two kinds of polypyrrole (PPy) films doped with two different dopants, i.e., dodecylbenzene sulfonate (DBS⁻) and p-phenol sulfonate (PPS⁻), on a silicon substrate using micromachining technology and electrochemical deposition of PPy. The fabricated actuators was 0.5mm long, 0.2mm wide and 1μm thick. In the fabrication of the microactuator, PPy(PPS) and PPy(DBS) were successively deposited on a patterned Au electrode, a part of which adhered to the silicon substrate through an adhesion layer of Cr.

The fabricated microactuator was actuated in an aqueous solution of sodium hexafluorophosphate (1mol/l) by applying a voltage of +0.7V to PPys versus a counter electrode (Pt). In the first voltage application, the Au layer underneath the PPy layers peeled away from the silicon substrate except the contacting part with the Cr layer and bent about 90 degree in about 18s, since adhesion force between Au and Si is enough smaller than that between Au and Cr. After the second voltage application, the microactuator bent stably having a response speed of about 5s.

INTRODUCTION

Conjugated polymers such as polyaniline, polypyrrole, and polythiophene, which change in volume by the insertion and the release of dopant in aqueous media, are excellent actuator materials, since they have large electrochemical strain (up to ~25%) and large stress (up to ~20MPa) under low application voltage (1-2V) [1-3]. Recently, microactuators using conjugated polymers, which are fabricated by combining with MEMS (Micro Electro Mechanical Systems) technology, are attracting much attention, aiming at realization of micromanipulator [4], microrobot arm [5], micro-valve, micro-pump etc.

One of the most stable conjugated polymers is polypyrrole (PPy). PPy is suitable for micro actuators, since it can be synthesized onto electrodes by electrodeposition from aqueous electrolyte containing pyrrole. Therefore, once micro electrodes are formed by micromachining technology, micro bimorph actuators, which bend by the difference in expansion of PPy and the electrode material, can be fabricated by depositing PPy electrochemically onto the electrodes.

In the previous works, gold was used as the electrode material [4,5]. In general, adhesion force between metal including gold and PPy deposited on the metal is not so good, i.e., PPy easily peels away from the metal. So, from a view point of the durability of the actuator, bilayer actuator of two kinds of different PPy is better.

Expansion rate of PPy to applied voltage is controlled by changing the kind of acceptor (dopant) doped into PPy. Therefore, we fabricated a bilayer microactuator of two kinds of PPys prepared with dodecylbenzene sulfonic acid (DBS) and p-phenol sulfonic acid (PPS) as supporting electrolyte for electrodeposition, on a silicon substrate.

EXPERIMENT

Fabrication of bimorph actuator

Before fabricating micro actuator, in order to evaluate bending characteristics of bimorph PPy actuator, a free-standing bimorph actuator was fabricated by electrochemically polymerizing PPy doped with dodecylbenzene sulfonate (DBS $^-$) (PPy(DBS)) and PPy doped with p-phenol sulfonate (PPS $^-$) (PPy(PPS)) successively onto a Ti plate. Polymerization was performed in an one-compartment three-electrode cell at 8°C by a current density of 1mA/cm^2. PPy(DBS) and PPy(PPS) were polymerized from an aqueous solution of pyrrole (0.5mol/l) and DBS (0.2mol/l), and from an aqueous solution of pyrrole (0.5mol/l) and PPS (0.2mol/l), respectively. The thickness of each PPy film was 10μm. After the polymerization, the deposited PPy(PPS)/PPy(DBS) film was peeled away from the Ti plate.

Fabrication of microactuator

Figure 1 shows a fabrication process of the PPy(DBS)/(PPy(PPS) bilayer microactuator. At first, a patterned Cr layer (thickness: 5nm), which had a function of an adhesion layer to fix the microactuator on the silicon substrate, was formed (figure 1(a)). Then, a Au layer (thickness: 100nm) was deposited on allover the substrate (figure 1(b)). After that, a photoresist layer, which had an opening having the shape of the microactuator (rectangle), was formed (figure 1(c)). Then, PPy(PPS) and PPy(DBS) were potentiostatically deposited successively on the Au layer in the opening of the photoresist in an aqueous solution of pyrrole (0.25mol/l) and PPS (0.15mol/l), and

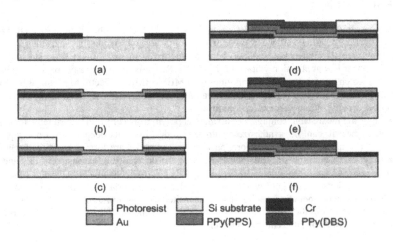

| Photoresist | Si substrate | Cr |
| Au | PPy(PPS) | PPy(DBS) |

56

in an aqueous solution of pyrrole (0.25mol/l) and DBS (0.15mol/l), respectively (figure 1(d)). After removing the photoresist layer (figure 1(e)), the Au layer except the part covered by the PPy bilayer was etched away (figure 1(f)). The fabricated micro actuator was 0.5mm long, 0.2mm wide and 1μm thick.

Measurement of bending characteristics

The bending characteristics of the bimorph actuator were evaluated in an aqueous solution of sodium hexafluorophosphate (NaPF$_6$)(1mol/l), and also in an aqueous solution of sodium chloride (NaCl)(1mol/l). The bending angle and the response speed of the bimorph actuator were evaluated by applying voltages (+0.7 to +1.5V) to the actuator versus a counter electrode (Pt).

The bending angle and response speed of the microactuator were also evaluated at a voltage of +0.7V versus a counter electrode (Pt) in an aqueous solution of NaPF$_6$ (1mol/l).

RESULTS AND DISCUSSION

Bending characteristics of bimorph actuator

Figure 2 shows photographs showing the bending of the bimorph actuator of PPy(DBS) and PPy(PPS), when voltages of ±0.7V were applied to the actuator versus the counter electrode (Pt) in the aqueous solution of NaPF$_6$. The actuator bended to the both sides according to the polarity of the voltage. The bending angle of about 90 degrees was achieved in a response time of 5s. This bi-directional bending is explained by the different bending behavior PPy(DBS) and PPy(PPS), i.e., PPy(DBS) and PPy(PPS) exhibit cation expansion and anion expantion, respectively [6]. Therefore, each PPy would stretch by the application of the voltage with the opposite polarity.

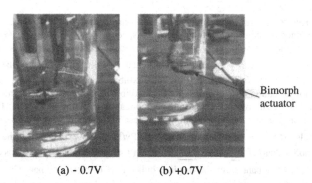

Bimorph actuator

(a) - 0.7V   (b) +0.7V

Figure 2. Photographs showing the bending of the bimorph actuator when voltages of ±0.7V were applied to the actuator versus the counter electrode (Pt).

Figure 3 shows effects of the applied voltage on the response time of the bending, when the voltage was applied in the aqueous solution of NaCl (1mol/l). The response time was measured at voltages between +0.7V and +1.5V by increasing the voltage from +0.7V by a step of 0.1V. After the measurement at +1.7V, the measurement started again by increasing the voltage from +0.7V. This cycle was repeated 5 times. In the first cycle, the response time decreased from 5.1s to 0.7s as the voltage increased. With increasing number of cycles, the response time decreased and approached 0.7s at all application voltages. As for bending angle, it decreased from about 90 degrees in the first cycle to about 40 degrees with increase in number of cycles.

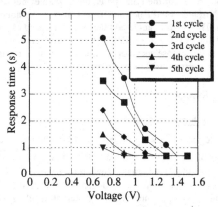

Figure 3. Effects of applied voltage on the response time of the bending, when the voltage was applied in the aqueous solution of NaCl (1/mol/l). Measurement was performed by changing the voltage from +0.7V to +1.5V by a step of 0.1V. These measurement were repeated 5 times.

The reason of these phenomena is considered to be as follows. As for the voltage dependence of the response time, since the speed of insertion and release of dopant is larger at higher application voltage, the response time would decrease with increasing application voltage.

The bimorph PPy actuator as fabricated contained PPS^- and DBS^- as dopant. It was reported that, in NaCl slution, PPy/DBS exhibited cationic expansion, i.e., it expands by the insertion of cation (Na^+), while PPy/PPS exhibited both cationic and anionic expansion during the initial cycling of cyclic step voltage application between –0.9V and +0.7V, and the cationic expansion decreased together with the increase in the anionic expansion with increasing number of cycles [6]. This behavior of strain of PPy/PPS was explained by the gradual anion (Cl^-) insertion instead of the cation (Na^+) insertion with potential cycling [6], as a result of the gradual exchange of dopant from the original bulky PPS to more compact Cl^-. The decrease in the response time, i.e., the increase in the response speed with the potential cycling is considered to come from increase in insertion speed of Cl^- due to the gradual dissipation of PPS into the solution.

Fabrication result of PPy(DBS)/(PPy(PPS) bilayer microactuator

Figure 4 is micro photographs of the micro actuator during the fabrication. Figures 4(a) and 4(b) are micro photographs after the removal of the photoresist layer (figure 1(e)) and after the etching of the Au layer(figure 1(f)), respectively. Figure 4(b) shows a neighborhood of the root of the microactuator.

The fabricated microactuator was actuated in the aqueous solution of $NaPF_6$ (1mol/l) by applying a voltage of +0.7V to the Au layer versus the counter electrode (Pt). In the first voltage application, the Au layer underneath the PPy(DBS)/(PPy(PPS) bilayer, peeled away from the silicon substrate except the contacting part with the Cr layer, and bent about 90 degrees in about 18s, since adhesion force between Au and Si is enough smaller than that between Au and Cr. Figure 5 shows a micro photograph near the tip of the micro actuator in bending, which was

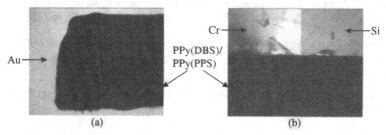

(a)   (b)

Figure 4. Micro photographs of the micro actuator: (a) after the removal of the photoresist layer (figure 1(e)), and (b) after the etching of the Au layer (figure 1(f)).

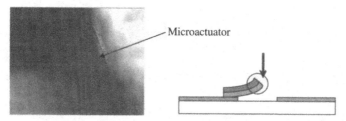

Figure 5. A micro photograph near the tip of the micro actuator in bending.

taken from above of the microactuator. After the second voltage application, the microactuator bent stably having a response time of about 5s.

CONCLUSIONS

The response time of the freestanding bimorph actuator of PPy(DBS)/PPy(PPs) in an aqueous solution of NaCl, decreased with increase in magnitude of positive voltage applied to the actuator and the number of the voltage application. This is explained by the increase in the insertion speed of dopant with the applied voltage and the gradual exchange of dopant in PPy(PPS) from PPS⁻ to Cl⁻ with potential cycling.

The bilayer microactuator of two kinds of polypyrrole (PPy) films doped with DBS⁻ and PPS⁻, was fabricated on a silicon substrate using micromachining technology and electrochemical deposition of PPy. The fabricated actuators was 0.5mm long, 0.2mm wide and 1μm thick. The microactuator was actuated in an aqueous solution of $NaPF_6$ by applying a voltage of +0.7V to PPys. It bent stably having a response speed of about 5s.

REFERENCES

1. S. Hara, T. Zama, W. Takashima, and K. Kaneto, *Polymer Journal*, **36**, pp.933-936 (2004).
2. T. Zama, S. Hara, W. Takashima, and K. Kaneto, *Bull. Chem. Soc. Jpn.*, **77**, pp.1425-1426 (2004).
3. T. Zama, N. Tanaka, W. Takashima, and K. Kaneto, *Polymer Journal*, **38**, pp.669-677 (2006).
4. E.W.H. Jager, E. Smela and I. Lundström, *Science*, **288**, pp.2335-2338 (2000).
5. E.W.H. Jager, E. Smela and O. Inganäs, *Science*, **290**, pp.1540-1545 (2000).
6. W. Takashima, S.S. Pandey, M. Fuchiwaki, and K. Kaneto, *Jpn. J. Appl. Phys.*, **41**, pp.7532-7537 (2002).

Device Application II

Mater. Res. Soc. Symp. Proc. Vol. 1134 © 2009 Materials Research Society   1134-BB04-01

High Frequency Length Mode PVDF Behavior Over Temperature

Mitch Thompson[1], Minoru Toda[1], Melina Ciccarone[2]
[1]Measurement Specialties, Inc., 460 E. Swedesford Rd., Wayne, PA 19087 U.S.A.
[2]Virginia Polytechnic Institute and State University, Blacksburg, VA 24061 U.S.A.

ABSTRACT

In an extension of earlier work, the temperature dependent parameters of PVDF operating in the length mode have been measured at frequencies useful in air ultrasound (20kHz to 100kHz) over a -45°C to +65°C temperature range. The length mode resonance of PVDF strips of different lengths was excited by mechanically clamping samples at the mid point during dielectric impedance testing conducted in a desiccated thermal chamber. Material properties were extracted from the impedance magnitude and phase angle data at temperature, and the various sample lengths allowed a range of resonant frequencies to be studied. Overall results generally confirm the visco-elastic behavior of PVDF. Testing was conducted on samples with both thin sputtered metal electrodes (~60nm) and thicker elastomeric silver ink electrodes (~8μm) to assess the performance difference. Silver ink is preferred in production as sputtered metal has current density an other limitations, but it causes a serious loss in performance at high frequencies.

INTRODUCTION

Piezoelectric PVDF has been successfully commercialized in a wide variety of applications, including: accelerometers, acoustic pickups, contact microphones, flexible switches, hydrophones, impact/shock sensors, medical vital signs, motion sensors, pyroelectric or thermal sensors, piezo cable, shock gages, dynamic strain gages, speakers, traffic sensors, and ultrasonic transducers. Because it is produced as a thin film, and because of its high uniaxial strain sensitivity in plane, PVDF is used in a length extensional mode in the vast majority of these applications. In all but a few of these length mode applications, PVDF is used as a sensor to convert elastic strain into a measurement voltage or charge over frequencies ranging from 0.1Hz to a few kHz and temperatures ranging from -40°C to +125°C.

As a result, piezo film performance has been reasonably well studied in this lower frequency bandwidth[1]. In addition, the thickness resonance mode behavior common for many ferroelectric materials has led to investigations into the behavior of PVDF in the MHz region[2-5]. Material parameters for PVDF in the range of 20kHz-100 kHz, however, have not been widely reported. Several air ultrasound applications operating in this general frequency range have become commercially successful in the last several years: digitizers and parametric speakers.

The digitizer products use PVDF transducers for both the transmission and reception of ultrasound pulses. When combined with an infrared LED timing signal, a pen or whiteboard marker having a PVDF transducer at its tip allows a module with multiple receivers to triangulate the position of the tip using the time of flight of an ultrasound pulse through the air. The typical operating bandwidth is in the 20kHz to 100kHz range and is the result of the compromise between increased attenuation and increased accuracy with increasing frequency.

A second emerging commercial application in this frequency range is the flat panel parametric speaker. These devices use a large area scalloped or corrugated PVDF film transducer driven at roughly 50kHz to create sound in an audio bandwidth through the generation of high intensity ultrasonic energy. The resulting audio output has a sharp directivity created by the combination of the ultrasonic carrier frequency and the relatively large aperture.

The proper design of these devices and others operating in this ultrasonic bandwidth requires material property data covering the proper temperature and frequency ranges. Length mode behavior is commonly measured using a device that applies an oscillating mechanical stress to the length of a piezo film sample[1]. This approach is generally limited to frequencies below several kHz due to resonances caused by the mass and compliance of the system.

The two most common electrode materials used in the production of commercial PVDF transducers are thin sputtered or evaporated metals and conductive inks. Printable conductive inks are available in a wide variety of compositions and use many different conducting agents, but a highly flexible silver filled urethane ink was used for this testing. Compared to thin metal electrodes, printed inks are substantially thicker, have much lower sheet resistivity, have much higher current density capability, and are far more robust mechanically. Further, it is much less expensive to produce complex electrode patterns using a printing process as opposed to the masked deposition or photolithography required to pattern metals. Because of these benefits, printed inks make are used as electrodes on the vast majority of commercial PVDF transducers. This work investigates how these inks affect PVDF in the mid ultrasonic frequency range.

EXPERIMENT

The measurement technique used to extract material parameter information from the impedance data is outlined in [6]. The measurements were taken using a computer controlled HP4192A impedance analyzer and thermal chamber. Custom fixturing was used to contact the film sample precisely to make electrical contact to the surface electrodes and to create the proper center clamped mechanical boundary conditions. All data is measured at the length mode resonance frequency.

Testing was done using 60°C annealed PVDF film with two different commercial electrodes: 150Å/450Å Ni/Al (NA), and a silver filled elastomeric ink (Ag) 8μm thick on both sides. PVDF density is about 1,880 kg/m^3 and the ink has a density of roughly 4,000 kg/m^3. Widths for these samples averaged about 1.7mm. The measurement technique is sensitive to noise, especially at the lower temperatures, and a number of samples of each kind were measured and the values averaged. Both the silver ink and sputtered metal samples were taken from adjacent sections of the same sheet of film.

RESULTS AND DISCUSSION

PVDF is visco-elastic and its material properties are a function of temperature and frequency. It is considered to be thermo-rheologically simple (TRS), and therefore Time-Temperature Superposition (TTS) can be applied[7]. Figures below show material properties for PVDF over a -45°C to +65°C temperature range. It should be noted that these plots represent the projections of three dimensional data onto the parameter and temperature or frequency axes.

Figure 1 shows the electro-mechanical coupling coefficient, k_{31}, of PVDF versus temperature. Each curve represents a sample (NA or Ag) of a different length (15mm or 60mm). The resonant frequency of each sample changes with temperature, as shown in Figure 4. Higher frequency is associated with shorter sample length, and this increase in measurement frequency causes the material to behave as if it were at a lower temperature. The spreading or separation in each pair of curves (NA and Ag) is caused by this frequency shift, and the shorter samples therefore undergo a k_{31} transition at higher temperatures. This is critical design information for air acoustic transducers that must operate over a wide temperature range.

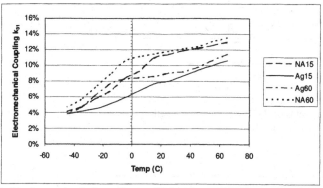

Figure 1. Electro-mechanical coupling coefficient, k_{31}, of PVDF samples with length of 15mm and 60mm having sputtered Ni/Al (NA) and Ag ink (Ag) electrodes.

The detrimental effect of the silver ink electrodes (Ag) is readily apparent. At room temperature an acoustic system depending on k_{31}^2 will show roughly 40% less sensitivity when using the silver ink electrodes.

Figures 2 and 3 show the mechanical Q and its inverse the mechanical loss tangent δ_M. The δ_M curves for the NA samples clearly show the PVDF glass transition event. The NA15 curve is shifted upwards in temperature compared to the NA60 curves because the shorter samples have higher resonance frequencies at each temperature. The range of glass transition temperatures for the NA samples, located at the peal of the loss term, lies between -15°C and +5°C, much higher than the -40°C often quoted for PVDF. The fact that the transition shifts higher into the commercial temperature range (-40°C to +85°C) highlights the motivation for this investigation.

The silver ink electrodes have a significant effect on the mechanical lossiness of the transducer. At lower temperatures and increased frequencies the ink loss term decreases. As the temperature rises above 25°C and the associated measurement frequencies decrease, the ink undergoes a transition and exhibits a dramatic increase in lossiness. This has a substantial effect on the sample Q as seen in the figure.

The sensitivity of transducers used in the air ultrasound applications noted above is affected by mechanical Q, particularly the parametric speakers which operate in a CW mode. The resonance frequency of a transducer is also significantly affected by the additional mass of the ink, as clearly shown in Figure 4. It should be noted that the output pressure of an ultrasound transmitter is proportional to the operating frequency.

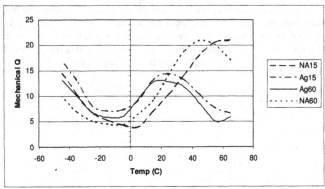

Figure 2. Mechanical Q of PVDF samples with length of 15mm and 60mm having sputtered Ni/Al (NA) and Ag ink (Ag) electrodes.

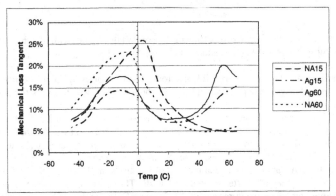

Figure 3. Mechanical loss tangent of PVDF samples with length of 15mm and 60mm having sputtered Ni/Al (NA) and Ag ink (Ag) electrodes.

The piezoelectric length mode strain constant d_{31} is shown versus temperature in Figure 5. The shorter NA samples here undergo a transition at a higher temperature than the longer NA samples, as expected because of the higher resonance frequencies. This is a clear indication of the TRS nature of PVDF – the higher frequency NA15 curve is essentially the same as the NA60 curve only shifted to the right, towards higher temperature. The fit to a TTS shift is not exact because the difference in resonant frequency is a function of temperature.

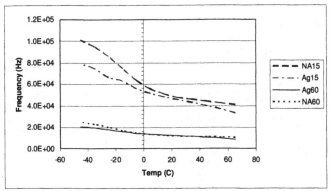

Figure 4. Measurement frequency for PVDF samples with length of 15mm and 60mm having sputtered Ni/Al (NA) and Ag ink (Ag) electrodes.

The silver ink electrodes cause a substantial decrease in piezo-activity over the bulk of the temperature range. The shift in the transition associated with the shorter, higher frequency sample is evident with ink electrodes as well. The transition temperature range is roughly the same for both electrode systems. The silver ink resonance frequencies are lower, but not low enough to shift to the curves to the left compared to the NA curves.

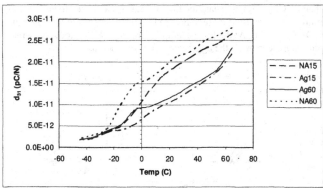

Figure 5. d_{31} of PVDF samples with length of 15mm and 60mm having sputtered Ni/Al (NA) and Ag ink (Ag) electrodes.

The d_{31} measurements of Figure 5 are shown against frequency in Figure 6. For each sample, the resonance frequency changed by roughly a factor of two over the temperature range. This behavior must be accommodated when designing resonant structures operating at ultrasonic frequencies over a broad temperature range. This is especially true for transducers with lower Q/ narrower bandwidth.

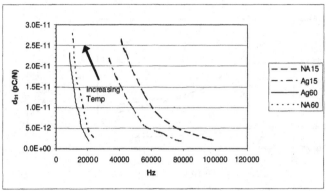

Figure 6. d_{31} of PVDF samples with length of 15mm and 60mm having sputtered Ni/Al (NA) and Ag ink (Ag) electrodes.

Young's modulus was calculated for each run and the results are given in Figure 7. The NA samples show the expected shift in the transition region, with the shorter samples showing an increased modulus at higher temperature as indicated by the arrows. This glass transition shift essentially matches the shift in the peaks of the δ_M curves as shown in Figure 3.

The Young's modulus of the silver ink samples in Figure 7 is distinctly higher than that of the NA samples, indicating that at ultrasonic frequencies the "soft" urethane based ink has a higher modulus than PVDF. From this data the ink modulus is calculated to be about 8GPa. This is counterintuitive to those using these inks at low frequencies, where they are much more compliant that PVDF film itself. At the lowest temperatures the PVDF modulus increases while the ink modulus does not, decreasing the aggregate effective modulus.

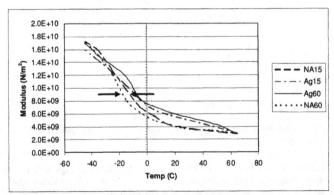

Figure 7. Young's modulus of PVDF samples with length of 15mm and 60mm having sputtered Ni/Al (NA) and Ag ink (Ag) electrodes. Arrows indicate transition region.

CONCLUSIONS

Elastic and piezoelectric material properties of PVDF have been explored in the 10kHz to 100kHz frequency range. Measurements indicate that the glass transition event is strongly influenced by temperature and frequency, confirming the visco-elastic nature of PVDF. The glass transition is that temperature/frequency at which PVDF undergoes a dramatic decrease in activity, a rapid increase in storage modulus, and a peak in loss modulus. Ultrasonic transducers operating in a useful temperature range must be designed to accommodate these frequency and temperature dependent properties.

The measurements herein have not been formally compared to behavior expected in thermo-rheologically simple polymers, but the PVDF samples with sputtered metal electrodes are generally consistent with TRS performance.

Although the silver filled elastomeric ink system used in this study offers many advantages as an electrode material for flexible, low frequency PVDF transducers, it has a detrimental effect on transducer performance in the ultrasonic frequency range. The ink has a modulus higher than that of PVDF at temperatures above about -20°C. This is a surprising finding and it means, unlike with low frequency applications, that the elastic effects of ink cannot be ignored when designing ultrasound devices. Further, the inks appear to be elastically complex materials with a mechanical lossiness which is a strong function of temperature.

REFERENCES

1. T.T. Wang, J. M. Herbert and A. M. Glass, Application of Ferroelectric Polymers. Blackie and Son Ltd., London. US edition by Chapman and Hall, New York, NY 1988
2. L. Brown and D. Carlson, "Ultrasound Transducer Models for Piezoelectric Polymer Films", IEEE Trans.Ultrason. Ferroelec. and Freq.Control, vol .36, pp. 313- 317 1989
3. M. Toda, "Cylindrical PVDF Film Transmitters and Receivers for air Ultrasound" IEEE Trans. Ultrason. Ferroelec. Freq Control. vol.49 pp. 626-633, 2002
4. M. Toda, "Phase Matched Air ultrasonic Transducers using Corrugated PVDF film with Half Wavelength Depth", IEEE Trans. Ultrason, Ferroelec. Freq. Control., vol.48, pp1568 -1574, 2001
5. Warren P. Mason, Physical acoustics vol. 1 -Part A pp.233-256 by D.A.Berlincourt, D.R.Curran and H.Jffe 1964, Academic Press, New York and London
6. M. Toda and M. Thompson, "Temperature Dependence of High Frequency Parameters of PVDF for Length mode Ultrasonic Air Transducers", Proceedings, IEEE Sensors Conference, Atlanta, GA, pp484-487, 2007
7. A. Vinogradov and F. Holloway, "Electro-mechanical Properties of the Piezoelectric Polymer PVDF", Ferroelectrics, Vol. 226, pp. 169-181, 1999

Mater. Res. Soc. Symp. Proc. Vol. 1134 © 2009 Materials Research Society   1134-BB04-03

Membrane as high performance biosensor platform

Xu Lu,[1,2] Liling Fu,[1] Shaokang Li,[2] Zhuo Xu,[2] and Wei Ren,[2] Anxue Zhang,[1] and Z.-Y Cheng[1]
[1] Materials Research and Education Center, Auburn University, Auburn, AL 36849, USA
[2] Electronic Materials Research Laboratory, Xian Jiaotong University, Xi'an 710049, China

ABSTRACT

Piezoelectric membrane was recently introduced as a high performance platform for the development of biosensor. The sensor principle is based on the resonance frequency change due to mass load. Therefore, determination of the resonance frequency is the key. In this paper, the fabrication and characterization of square membranes made of piezo-PVDF thin film are reported. The thickness of the PVDF is about 30 μm, while the size of the membrane is from 2 mm to 6 mm. The resonance behavior of these membranes was characterized under different conditions: two sides with air in different pressure, one side in air and another side in different liquids. It is experimentally found that the membrane based devices, working in the bend mode, could keep a nearly same Q value in water than in air, which makes the piezoelectric membrane a stronger candidate for biosensor platform used in liquid environment. Furthermore, the influences of size, pressure difference, density and viscosity of the environments on the resonance frequency and the Q value are experimentally determined and the results are discussed.

INTRODUCTION

In recent years, analytical technologies with high sensitivity and short response time for detecting a small amount of deleterious particles are eagerly needed in many important areas of everyday life, such as food industry, medical diagnostic system, environment mastery and public safety. The acoustic wave (AW) devices are considered as strong candidate for biosensor platform, since the directness, immediateness, high sensitivity and low cost. Mass sensitivity (S_m) and Q value are considered as the most important parameters of the AW biosensors. The S_m indicates the frequency shift due to a unit mass load and the Q factor describes the sharpness of the resonance peak. Micro-cantilever has been widely investigated as biosensor platform since the high mass sensitivity. [1-2] However, there are still some problems that limit the real application of this device, most important, small Q value in liquids. [3-4] Since most of micro-organisms exist in liquid environments, detection methods that could be directly carried out in liquid are badly desired. Diaphragm based biosensor, which could exhibit a nearly same Q value both in air and in liquid has been reported. [5] In this paper, the piezoelectric square membrane is introduced as high performance biosensor platform. To determine the behavior, the membranes were tested under different conditions: two sides with air in different pressure, one side in air and the other side in different liquids. The influences of size, pressure difference, density and viscosity of the environments on the resonance frequency and Q value have been investigated. And results tell that the membrane could be considered as strong candidate for biosensor platform, especially working in liquid environments.

EXPERIMENT

Membranes made of commercial piezoelectric polyvinylidene fluoride (PVDF) thin film are used in the experiments. The PVDF film (30 μm in thickness) was coated with square gold electrodes (100 nm in thickness) by sputtering method and then cemented on the PMMA holders (3mm in thickness and with square hole in different sizes from 2mm*2mm to 6mm*6mm) as a sandwich structure and fixed with screws. One side of the membrane was bonded with a PMMA spigot and connected to a micro pump, while the other side is open to air. The micro pump is used to bring pressure difference on both sides of the membrane, which would dome the film and induce stress as restore force of the vibration. The structure of the device is shown in figure 1.

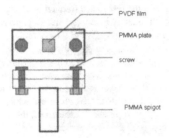

PVDF film

PMMA plate

screw

PMMA spigot

Figure 1. The schematic structure of the PVDF membrane

To determine the performances, the membranes were tested both in air and in liquids. For the air environment tests, the resonance behaviors were recorded under difference pressure differences by controlling the pumping rate. For the liquid environment tests, the pressure difference were fixed at 650 Torr, and membranes were tested in different liquids by filling the cylinder formed by the hole in the PMMA holders and the membrane as the bottom. Commercial reference oils from SavanTech and some kinds of common liquids were used in the liquid tests. The viscosities and densities of air and the liquids are given in table I.

Table I. Viscosities and densities of air and the liquids used in the experiments.

Liquid	Air	Hexane	Ethanol	2-Propanol	4-Methyl-2-Pentanol	1-Octanol	R-100	R-300	R-2350	R-2450	R-600
Viscosity (cP)	$1.79*10^{-2}$	0.30	1.70	2.04	4.07	7.29	19.88	61.94	152.51	372.66	1733
Density (g/cm³)	$1.29*10^{-3}$	0.6548	0.7914	0.7855	0.8075	0.8262	8.8292	0.8527	0.8608	0.8700	0.8863

Both impedance and phase signals are recorded by Agilent 4294A precision impedance analyzer. The resonance frequency f_0 is considered as the frequency when phase gets its peak, and Q factor is considered as $f_0/\Delta f$, where Δf is the band width.

RESULTS AND DISCUSSION

The theoretic resonance frequency of membrane is given as: [6]

$$f_{ij} = \frac{\alpha_{ij}}{2a}\sqrt{\frac{T}{\rho h}} \qquad i,j = 1,2,3...$$

(1)

Where α_{ij} is the dimensionless parameter, T, h, ρ and a are the tension, thickness, density and side length of the membrane respectively.

The experimental resonance frequencies of membranes under difference pressure differences are show in figure 2. It is clear that the frequency is approximately linearly dependent on the a^{-1} as expected by equation 1, which means that the devices working as membrane.

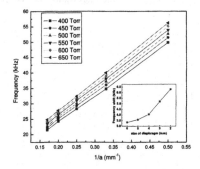

Figure 2. Resonance frequency VS a^{-1} under different pressure differences (where a is the sidelength of the membrane). Insert figure: Frequency shift vs the side length of membrane
A topical resonance spectrum of membrane is shown in figure 3.

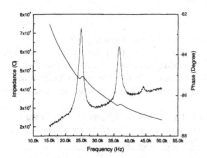

Figure 3. Resonance behavior of 6mm*6mm membrane tested under a pressure about 650 Torr in air.

The frequencies of different peaks are marked and consulted with the α_{ij}, which restricts the frequency rates of different modes, as shown in table II.

The mode orders of those peaks could be identified: the first peak is the first mode; the second peak is the second mode, and so on.

Table II. The α_{ij}, and experimental resonance frequencies of a 6mm*6mm membrane tested under a pressure difference about 650 Torr.

Mode Sequence	1	2	3	...
α_{ij}	1.414(11)	2.236(12, 21)	2.828(22)	...
Experimental Frequency (kHz)	24.78	36.66	44.21	...

The frequency shifts of different modes of a 6mm*6mm membrane are shown in figure 4. As pressure difference increasing, the frequency of each mode shifts to higher range. And the frequency shift gets bigger when the mode order gets higher. It also can be perceived that the shift of resonance frequency gets smaller while the size gets bigger, as the inert figure showed in figure 2.

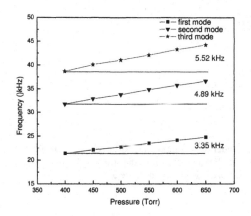

Figure 4. Frequency shifts of different modes of 6mm*6mm membrane

It is easy to understand the frequency shift of each mode. As the pressure difference gets bigger, the stress in the membrane gets bigger too. According to equation 1, the frequency gets bigger as an indirect result of the increasing pressure difference. While the pressure difference influences on different modes and different sizes may because of the dome shape induced by the pressure difference.

Q values of different modes are shown in table III. Higher modes have higher Q values. And when the membrane works in the third mode, a Q value more than 40 has been got.

Table III. Q values of different modes

Mode order	Q value 2mm	3mm	4mm	5mm	6mm
1	13.1	15.4	15.6	14.4	14.6
2		24.3	24.5	25.5	28.1
3			43.1	42.7	46.8

The Q values of membranes tested in air and different liquids are shown in figure 5.

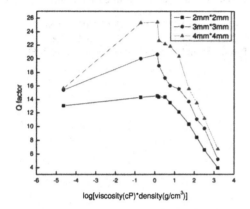

log[viscosity(cP)*density(g/cm³)]

Figure 5. Q values of membranes tested in liquids of different densities and viscosities.
It is found that when the density*viscosity is between $2.3*10^{-5}$cP*g/cm³ (air) and
1.35cP*g/cm³ (Ethanol), Q value is getting bigger. And Q value is decreasing when
density*viscosity increasing from 1.35cP*g/cm³ to 1536cP*g/cm³. As the densities of those
liquids are not far away from 1g/cm³, the Q factor is influenced much by viscosity than density in
this experiment.

The Q value is not changing much around 1 cP*1g/cm³, which is almost the viscosity and
density of the liquids in real detection and the results means that the membrane is a stronger
candidate for biosensor working in liquid environment.

CONCLUSIONS

This paper describes a novel biosensor platform based on membrane structure. The
resonance behavior of the membrane has been investigated both in air and liquid. The influences
of size, pressure, density and viscosity on the resonance frequency and Q value have been
discussed. The experimental results show that the membrane can be considered as a strong
candidate for biosensor platform used in liquid environments.

ACKNOWLEDGMENTS

This research is supported by the China Scholarship Council [2007]3020 and NFSC
(Grant No. 60528008).

REFERENCES

1.   R. Raiteri, M. Grattarola, H. J. Butt and P. Skládal. Sens. Actuators. B **79**, 115 (2001)

2. C. Ziegler. Anal. Bioanal. Chem. **379**, 946 (2004)
3. N. V. Lavirk, M. J. Sepaniak and P. G. Datskos. Review of Scientific Instruments. **75**, 2229 (2004)
4. S. Q. Li, L. Qrona, Z. M. Li and Z.-Y. Cheng. Appl. Phys. Lett. **88**, 073507 (2006)
5. X.Yang, Z. M. Li, L. Odum and Z.-Y. Cheng. Appl. Phys. Lett. **89**, 223508 (2006)
6. R. D. Blevins, Formula for Natureal Frequency and Mode Shape (Krieger, Malabar, FL, 1979) p. 226

Mater. Res. Soc. Symp. Proc. Vol. 1134 © 2009 Materials Research Society 1134-BB04-09

Two-Way Shape Memory Polymer Composite and Its Application

Woong-Ryeol Yu[1], Seok Jin Hong[1] and Ji Ho Youk[2]
[1]Materials Science and Engineering, Seoul National University, Seoul, South Korea
[2]Advanced Fiber Engineering, Inha University, Incheon, South Korea

ABSTRACT

Thermo-responsive shape memory polymer is a smart material that turns a deformed (or temporary) configuration into the permanent (or original) one by external heat. The temporary shape can be processed using the mechanical forces above a certain temperature, which is called the transition temperature (T_m or T_g), and maintained until the material gains the thermal energy enough to reach its transition temperature. Thereafter, it will recover the memorized permanent shape according to heating above the transition temperature. There are two possibilities for subsequent cooling. As the material loses thermal energy, its shape will change to another permanent one or nothing will happen. The former case is called as two-way shape memory (TWSM) behavior while the latter is one-way shape memory (OWSM) one. In this study, we report that TWSM behavior can be realized in OWSM polymer using a modified thermo-mechanical process.

INTRODUCTION

Shape memory polymers (SMPs), which are able to restore the memorized permanent shape in response to external stimuli, have attracted much interest as a smart material [1-4]. SMPs have advantages such as low cost, low density, high recoverable deformation of more than 200%, and easy processing over shape memory alloys (SMAs) [5, 6], and various potential applications, e.g., sensor, actuator, muscle, biodegradable sutures, catheters, smart stent, etc, have been reported [7-9].

The characteristics of SMP can be explained by its thermo-mechanical behavior in Figure 1. These shape memory behavior can be generally represented by using transition temperature (T_{trans}), such as T_g and T_m of soft-segment in Polyurethane. First three steps (a-b-c-d path) are to process a temporary shape. After these steps, the SMPs recover the permanent shape from the temporary one on heating as shown in Figure 1 (d-e path). To invoke the shape memory behavior again, the SMPs need to be processed again to temporary shapes using the same procedure (a-b-c-d path). Since the spontaneous shape change occurs only upon heating after forming SMPs into the temporary shapes, such behavior is called as one-way shape memory (OWSM) one.

Shape memory polymers like liquid crystal elastomers (LCEs) can maintain the anisotropic nematic phase in low temperature whereas the isotropic phase is formed in high temperature. As a result, LCEs show two memorized shapes, i.e., upon heating they deform to one configuration, and subsequent cooling causes them to change macroscopically to the other configuration beyond the thermal contraction. This reversible and spontaneous behavior is called two-way shape memory (TWSM) behavior [10].

This paper will report that thermo-responsive OWSM polymers can be tailored to show the TWSM behavior, that is, on heating they deform to one shape and on cooling to the other shape in a repeatable and spontaneous manner.

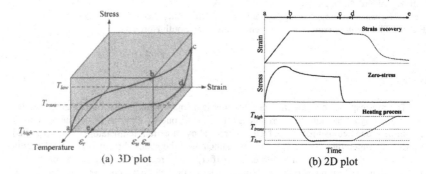

(a) 3D plot   (b) 2D plot

Figure 1. Thermo-mechanical behavior of SMPs. (a-b path): deforming a SMP at T_{high}. (b-c path): cooling to T_{low}, keeping the strain fixed. (c-d path): unloading to zero stress. (d-e path): heating to T_{high}, causing it to recover its permanent shape. (a-b-c-d path): processing of the temporary shape, (d-e path): recovering of the memorized permanent shape.

EXPERIMENTAL

Two polyurethanes were synthesized using poly (ε-caprolactone) (PCL), 4,4'-diphenylmethylene diisocyanate (MDI), 1,4-butanediol (BD) as a chain extender by one-step process. The molar ratios of MDI/PCL/BD were 6/1(Mn:4000)/5 and 2/1(Mn:3000)/1, which will be referred to sample I and II, respectively. Since the molecules in the soft segment of these polymers can form crystallites, their melting point can serve as the transition temperature (T_{trans}). Because T_{trans} is a temperature to show the spontaneous recovery of the deformation according to the temperature, it can be also measured by using the thermo-mechanical test (a-b-c-d-e path in Figure 1). T_{high} and T_{low} were set 80 and 0°C, respectively. The crosshead speed was set 20mm/min at T_{high} in a-b path, and a cooling rate of -40°C/min was maintained in b-c path. After the load release step (c-d path), the samples were heated from T_{low} to T_{high} at a heating rate of 5°C/min under zero load condition. Then, T_{trans} was determined by finding the central position of the recovery strain path between two plateaus.

A modified thermo-mechanical test was designed to investigate the shape memory behavior of the two samples, and focus on any possibility of reversible and spontaneous TWSM behavior of OWSM polymers (see Figure 2(a)). This test is very similar to the general thermo-mechanical test as shown in Figure 1, however a new step (d-f path) was introduced instead of d-e path. In other words, the sample is heated from T_{low} to T_{high} under the constant strain. This step causes increase of stress due to recovery of temporary shape under the strained strain. Then, the strain variation according to the cyclic temperature process is recorded under the constant stress (f-g path).

TWSM behavior of SMPs was recorded after temporary shaping process, employing the constant stress (a-b-c-d-f path in Figure 2). First, the rectangular sample was heated to T_{high}

($T_{trans}+20°C$) and elongated to 200% strain at a crosshead speed of 20mm/min (a-b path). Under the fixed strain condition, the strained sample was cooled to T_{low} ($T_{trans}-20°C$) at a cooling rate of about $-40°C/min$ (b-c path). As the load is released, a little shrinkage of length can be observed (c-d path). Then, the sample was heated from T_{low} ($T_{trans}-20°C$) to T_{high} ($T_{trans}+20°C$) at a heating rate of $5°C/min$ under the constant strain condition, showing the stress developed due to the constrained strain (d-f path). Repeated thermo-cycles between T_{low} ($T_{trans}-20°C$) and T_{high} ($T_{trans}+20°C$) were applied to the samples under the constant stress condition (f-g path).

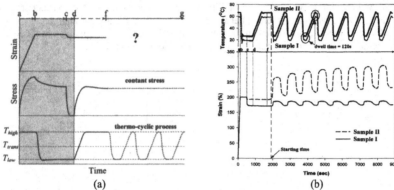

| (a) | (b) |

Figure 2. (a): Modified thermo-mechanical test. (b): Reversible and spontaneous strain variation (TWSM behavior) according to the cyclic temperature.

DISCUSSION

The transition temperatures of the samples I and II were measured using thermo-mechanical test as shown in Figure 3.

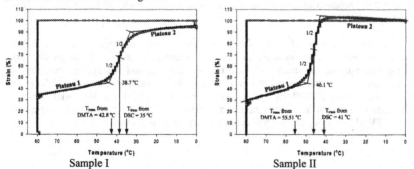

| Sample I | Sample II |

Figure 3. Determination of the T_{trans} using the thermo-mechanical test (T_{high}: 80°C and T_{low}: 0°C).

TWSM behaviors of samples I and II were investigated by the modified thermo-mechanical test. The dwell time between the cyclic temperatures was set to be 120 seconds.

Interestingly, the cyclic variation of strain was observed. The strain was increased on cooling and decreased on heating (see Figure 2(b)), implying that the SMPs show the TWSM behavior.

An additional thermo-mechanical test was carried out using sample II. The effect of the constant stress on the TWSM behavior was investigated by varying constant loads as shown in Figure 4(a). The dwell time was set to 120 seconds. As the constant stress increases, the amplitude of the cyclic strain variation was increased; however the creep behavior was observed for the constant stress of more than 0.29MPa. The effect of the dwell time on the repeatable and cyclic shape strain was also investigated with a dwell time of 30 minutes as shown in Figure 4(b). The TWSM behavior shows good repeatability with non-creep strain for the constant stress of less than 0.32MPa, whereas the creep behavior was manifested for the constant stress of more than 0.415MPa. Interestingly, the creep strain was only induced at T_{high}, implying that the rigidity of soft-segment is strong enough not to permit the creep deformation at T_{low}.

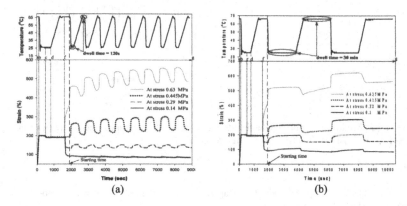

Figure 4. TWSM behavior of sample II under the constant stress condition: (a) dwell time=120s, (b) dwell time=30mins.

The creep behavior of SMPs should be avoided for their applications to thermo-responsive sensors; therefore an optimization of the constant stress is required for non-creep strain. For this purpose the strain difference between the first and last cycles (the difference between strain at starting time and last strain in Figure 4) was calculated (see Figure 5(a)). The optimized stress value (0.338MPa) was chosen from a set of 0.33~0.35MPa with non-creep strain and used for the cyclic thermo-mechanical test with dwell time of 120 second. Figure 5(b) shows that TWSM behavior can be obtained without the creep deformation.

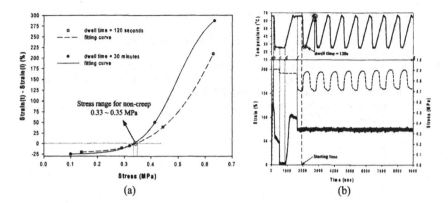

(a)

(b)

Figure 5. Optimization of TWSM behavior without the creep strain. (a) strain difference between the first and last thermo-cycles. (b) TWSM behavior by the optimized stress.

CONCLUSIONS

We reported a new finding that can simply impart the TWSM behavior to shape memory polyurethanes using a thermo-mechanical treatment. The effect of the constant stress and the dwell time on the TWSM behavior was investigated, concluding that the current finding can be used in developing two-way thermo-responsive sensors from thermo-responsive polyurethanes through an optimization study.

ACKNOWLEDGMENTS

The authors would like to thank the Korea Science and Engineering Foundation (KOSEF) for sponsoring this research through the SRC/ERC program of MOST/KOSEF (R11-2005-065).

REFERENCES

1.   S. K. A. L. H.Y. Jiang, Advanced Materials 18 (11), 1471-1475 (2006).
2.   A. Lendlein, H. Jiang, O. Junger and R. Langer, Nature 434 (7035), 879-882 (2005).
3.   J. S. Leng, X. Lan, Y. J. Liu, S. Y. Du, W. M. Huang, N. Liu, S. J. Phee and Q. Yuan, Applied Physics Letters 92 (1), 014104 (2008).
4.   A. Matsuda, J. i. Sato, H. Yasunaga and Y. Osada, Macromolecules 27 (26), 7695-7698 (1994).
5.   C. Liu, H. Qin and P. T. Mather, Journal of Materials Chemistry 17 (16), 1543-1558 (2007).

6.   A. M. S. H. L. H. Y. Witold Sokolowski and R. Jean, Biomedical Materials **2** (1), S23 (2007).
7.   C. M. Y. Y. L. R. S. N. W. K. S. A. Ken Gall, Journal of Biomedical Materials Research Part A **73A** (3), 339-348 (2005).
8.   A. Lendlein and R. Langer, Science **296** (5573), 1673-1676 (2002).
9.   W. I. Small, M. F. Metzger, T. S. Wilson and D. J. Maitland, Selected Topics in Quantum Electronics, IEEE Journal of **11** (4), 892-901 (2005).
10. J. Naciri, A. Srinivasan, H. Jeon, N. Nikolov, P. Keller and B. R. Ratna, Macromolecules **36** (22), 8499-8505 (2003).

Poster Session:
New Materials and Devices

Mater. Res. Soc. Symp. Proc. Vol. 1134 © 2009 Materials Research Society   1134-BB05-01

Amplified Fluorescence Turn-on Assay for Mercury(II) Based on Conjugated Polyfluorene Derivatives and Nanospheres

Yusong Wang and Bin Liu*
Department of Chemical and Biomolecular Engineering, 4 Engineering Drive 4, National University of Singapore, Singapore 117576

ABSTRACT

Detection of mercury with high sensitivity and selectivity constitutes a significant research concern. Here, we report an amplified fluorescence turn-on assay for mercury(II) with an improved performance. This sensing system takes advantage of optically amplifying fluorescent conjugated polyfluorene derivatives and DNA immobilized silica nanospheres (NSs) in addition to the specific thymine- mercury(II)-thymine(T- Hg^{2+}-T) interaction. The employment of ion-specific T- Hg^{2+}-T coordination increases the melting temperature (T_m) of the double-stranded DNA (dsDNA) on the hybridized NS surface. After thermal washing at 45 °C, the Hg^{2+} treated sample (dsDNA-NS) was effectively differentiated from that treated with nonspecific ions through monitoring fluorescence emission of fluorescein (Fl) labeled target DNA remained on the NS surface. Finally, a cationic conjugated polyfluorene derivative (CCP) was introduced to electrostatically associate with the DNA molecules on the NS surface, resulting in an amplified Fl signal via fluorescence resonance energy transfer (FRET) from the CCP to the dye molecule. In comparison with the use of Fl alone as a signal reporter, the presence of CCP significantly enhances the detection fluorescence intensity, reduces false-positive signal, and improves the detection selectivity for mercury(II). Further improvement in the probe design could yield more efficient metal ion sensors, which have the potential to be operated at room temperature and for the detection of other metal ions besides mercury(II).

INTRODUCTION

Recognition and detection of mercury(II) (Hg^{2+}) and its derivatives is of vital importance due to their deleterious effect on the environment and human health [1]. Different strategies for the analysis of Hg^{2+} have been developed, and the majority is based on fluorescence and colorimetric methods. Recently, the complexation of metal ions with nucleotide purine and pyrimidine bases has attracted considerable interests [2], and the specific interaction between thymine-mercury(II)-thymine (T- Hg^{2+}-T) has been widely used for Hg^{2+} detection [3-6].

Water-soluble conjugated polymers (CPs) as optically sensitive materials have been widely used in chemical and biological sensors [7]. In comparison with small molecular counterparts, the large absorption cross section and delocalized electronic backbone structure of polymers allow efficient light-harvesting and rapid intrachain and interchain energy transfer [7-10]. Particularly, cationic conjugated polyfluorene derivatives (CCPs), have been proven useful for signal amplification of dye labeled biomolecules through fluorescence resonance energy transfer (FRET) from the CCPs to dye molecules attached to the biomolecules [11-12]. By taking advantage of the signal amplification of CCPs and the specific T-Hg^{2+}-T interaction onto the DNA immobilized silica nanospheres (NSs) surface, in the present study, we report an amplified fluorescence turn-on assay for mercury(II) detection. With the signal amplification provided by the CCP, the NS-based assay has shown an improved Hg^{2+} detection relative to the detection without the CCP.

EXPERIMENT

Materials:

The CCP was synthesized according to the previous report [13]. DNA oligonucleotides were purchased from Research Biolabs (Singapore). Tween-20, sodium nitrate, silver nitrate was purchased from (Sigma Aldrich). Other metal salts used in this work are: $Na(OAc)$, $Hg(OAc)_2$, $MgSO_4$, $Co(NO3)_2$, $Ni(OAc)_2$, $Ba(OAc)_2$, $K(OAC)$, $Pb(NO_3)_2$, and $Cd(SO_4)$. Phosphate buffered saline is a commercial product (ultrapure grade, 1st BASE). The prepared buffers include: the hybridization buffer containing 137 mM NaCl, 2.7 mM KCl, 10 mM phosphate and 0.1% Tween 20 (PH = 7.4), the acetate buffer containing 140 mM sodium nitrate and 10 mM sodium acetate (PH = 7.1), and the 25 mM PBST buffer containing 25 mM PBS with 0.1% Tween-20 (PH = 7.2). MilliQ water (18.2 MΩ) was used to prepare all buffer solutions. The stock solution of Hg^{2+} (2 mM) was prepared by dissolving 6.78 mg of $Hg(OAc)_2$ in 10.64 mL acetate buffer with few droplets of concentrated HNO_3. Other metal ion stock solutions (2 mM) were prepared by dissolving metal salts in the acetate buffer.

Characterization:

The concentration of DNA was determined by measuring the absorbance at 260nm using a Shimadzu UV-1700 spectrometer. Photoluminescence (PL) was measured using a Perkin Elmer LS-55 equipped with a xenon lamp excitation source and a Hamamatsu (Japan) 928 PMT, using 90 degree angle detection for solution samples. The as-prepared NSs were imaged at the edge of the dried samples (for fresh synthesized silica NSs) and the center of the dried samples (for oligonucleotide modified silica NSs), using a field emission scanning electron microscope (FE-SEM JEOL JSM-6700 F) after coating a Pt layer (~5 nm in thickness) via a platinum coater (JEOL JFC-1300, Auto Fine Coater). The sampling solution has the concentration of ~0.5 mg/mL, and a 1μL droplet is loaded onto a polished copper tape supported by a copper stub.

Mercury(II) detection procedure:

Hybridization of DNA-immobilized NSs. Bare silica NSs and amine-labeled DNA immobilized NSs were prepared according to our previous report [14]. The ssDNA-NSs were washed twice with the hybridization buffer before hybridization. Fl-labeled oligonucleotide was then added into the ssDNA-NS solution in hybridization buffer ([Fl-DNA] = 1 μM, [NS] = 0.8 mg/100 μL). The mixture was shaken at room temperature for 2 h. After incubation, the suspension was centrifuged, and the precipitate was washed with hybridization buffer while the supernatant was kept for fluorescence analysis. Finally, the hybridized dsDNA-NSs were dispersed in the hybridization buffer ([NS] = 0.1 mg/50 μL).

Ion incubation and thermal wash. The as prepared dsDNA-NS was transfered from the hybridization buffer to the acetate buffer, and packed with different [NS]. Hg^{2+} or the nonspecific ion mixture (including eight different ions, each has the same concentration as Hg^{2+}) in the acetate buffer was added into the dsDNA-NS solutions (3 μM for each metal ion). After an incubation at room temperature for 45 minutes with gentle shaking, the above solutions were centrifuged and the precipitates were dispersed in the acetate buffer ([NS] = 0.1 mg/500 μL acetate buffer).The metal ion treated samples were then loaded into an eppendorf thermomixer (45 °C, 1200 rpm, 15 minutes) for thermal wash. After that, the treated DNA-NSs were centrifuged immediately. The precipitates were washed with the hybridization buffer once, and then dispersed into the 25 mM PBST buffer for PL measurement.

Fluorescence measurement and analysis. 0.1 mg of the hybridized NSs (after metal ion incubation and thermal wash) was dispersed in a 25 mM PBST buffer in a quartz cuvette for fluorescence measurement. The solution fluorescence was measured upon excitation at 490 nm. The CCP was then added to the metal ion treated NS solution (final [CCP] = 0.2 μM), and the emission spectra were collected upon excitation of the CCP at 370 nm. For all the Fl emission spectra, the background emission and the polymer emission tail were subtracted unless specified.

DISCUSSION

Preparation of hybridized DNA-NS

The oligonucleotide modified silica NSs have ~370 NH_2-DNA molecules on each NS (Figure1). The number of DNA is calculated from the absorbance difference at 260 nm for the DNA solution before immobilization and the supernatant after immobilization, considering 1.0 mg of 100 nm NSs contain ~1 × 10^{12} number of NSs [14]. After hybridization and centrifugation, the extent of hybridization is estimated from the fluorescence difference between the Fl-DNA solution before hybridization and the supernatant after hybridization. The fluorescence intensity at 522 nm is observed as the maximum value in the Fl-DNA PL spectrum upon excitation at 490 nm, and used for fluorescence comparison. On the average, there is ~70 Fl-DNA molecules captured by a single ssDNA-NS, which corresponds to a hybridization efficiency of ~20 %.

Figure 1. FE-SEM images of the synthesized silica NSs with ~100 nm in diameter with a polydispersity of 6% based on calculating 150 NSs. The inset represents oligonucleotide modified silica NSs.

<u>**Sensing strategy**</u>

The full Hg^{2+} detection method is shown in Scheme 1. The assay begins with probe (NH$_2$-DNA, 5'-NH$_2$-GTGACCATTTTGCAGTG-3') immobilized silica NSs in solution. After hybridization with fluorescein (Fl) labeled DNA (Fl-DNA, 5'-Fl-CACTGCATTTTGGTCAC-3'), double-stranded DNA (dsDNA) containing three pairs of T-T mismatches is formed on the NS surface. The recognition is accomplished by Hg^{2+}-specific coordination between the T-T mismatches on the silica NSs (Hg^{2+}, shown in orange dot, non-specific ions, shown in blue square). In the presence of Hg^{2+}, T-Hg^{2+}-T complexation increases the melting temperature (T_m) of the resulting duplex, and Fl-DNA remains on the NS surface during thermal washing. In contrast, the non-specific ions can not form stable metal-DNA complex, and the duplex will denature during thermal washing, leaving weak or no fluorescent signals on the NS surface [15]. In the final step, addition of the CCP leads to amplified fluorescence signal of the Fl-DNA remained on the NS surface.

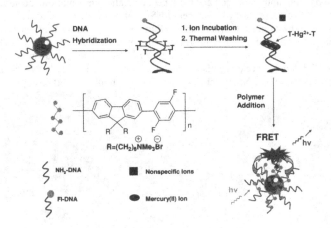

Scheme 1. Schematic representation of a CCP-amplified fluorescence turn-on assay for mercury(II).

The CCP has a fluorenephenylene backbone with a functionalization of the phenyl ring with fluorine (structure shown in Scheme 1), and emits blue in color. This functionalization allows for good spectral overlap and matched energy levels between the CCP and Fl so that efficient energy transfer could occur [13]. Monodispersed silica NSs (~ 100 nm in diameter, see Figure 1) were chosen to allow separation and washing steps. In addition, these well-defined NSs do not have interference on the fluorescence properties of CCPs. The synthesis and modification of NSs are according to the previous report [14]. The probe pair of 5'-Fl-CACTGCATTTTGGTCAC-3' and 5'-NH$_2$-GTGACCATTTTGCAGTG-3' is specially chosen to allow efficient hybridization and meanwhile a large difference in T_m for the duplex before and after Hg^{2+} treatment. According to the previous report [2b], the duplex with three pairs of T-T mismatch has a melting temperature of 43 °C, which increases to 61 °C upon treatment with excess Hg^{2+}. At 45 °C, there is no denaturation observed for the duplex containing three pairs of

88

T-Hg^{2+}-T, while the duplex with three T-T mismatches is almost completely denatured [2b]. This result allows us to conduct thermal wash at 45 °C to selectively remove the hybridized Fl-DNA molecules from the NS surface in the presence of non-specific ions.

Fluorescence mercury(II) assay using hybridized DNA-NS

Fig. 2 shows the fluorescence spectra for dsDNA-NS suspensions (0.1 mg NSs) after the NSs were treated with the metal ions, which was followed by thermal washing at 45 °C. The metal ions include Hg^{2+} and the nonspecific ion mixture, which contains Ag$^+$, Mg^{2+}, Co^{2+}, Ni^{2+}, Ba^{2+}, K$^+$, Pb^{2+}, and Cd^{2+} (3 μM for each metal ion). Measurement was conducted after dispersing the metal ion treated NSs in 25 mM PBST buffer.

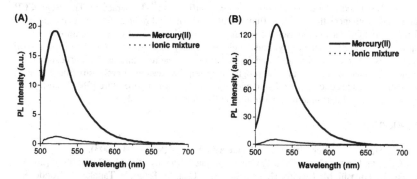

Figure 2 Mercury(II) sensing with the aid of CCP. **(A)** The Fl emission spectra of the samples treated by metal ion (Hg^{2+} or non-specific ion mixture, [Metal Ion] = 3 μM) upon excitation at 490 nm. **(B)** The CCP sensitized Fl emission spectra at [CCP] = 0.2 μM for the same samples upon excitation at 370 nm. Measurement was done in 25 mM PBST buffer. [NS] = 0.1 mg/mL.

As shown in Fig. 2A, Hg^{2+} treated sample shows higher PL intensity in comparison with that treated by the non-specific ion mixture. The maximum fluorescence intensity of Hg^{2+} treated sample was about 15-fold higher than that in the presence of non-specific ion mixture. Although comparison of the NS solution fluorescence intensity in the presence and absence of Hg^{2+} (shown in Fig. 2A) indicates the selectivity of the assay for Hg^{2+} detection, the absolute solution fluorescence intensity is rather low and easily affected by false-positive signals.

We then employed the CCP to improve the detection performance. The polymer sensitized Fl emission intensities at [CCP] = 0.2 μM for NS solutions treated with Hg^{2+} or the non-specific ion mixture (3 μM for each metal ion) are shown in Fig. 2B. The maximum fluorescence intensity of Hg^{2+} treated sample was over 25-fold higher than that in the presence of non-specific ion mixture. The low signal for non-specific ions is due to the limited number of Fl-DNA remained on the NS surface after thermal washing. When one compares the detection in the presence and absence of the CCP, there is an obvious improvement in detection selectivity when the CCP is used as the energy donor.

Moreover, when compared Figure 2A and Figure 2B, upon excitation of the polymer at 370 nm, Fl emission intensity was enhanced both for Hg^{2+} treated and nonspecific ion mixture treated samples, relative to the direct excitation at 490 nm. For the Hg^{2+} treated sample, excitation of the polymer ([CCP] = 0.2 μM) at 370 nm leads to a Fl emission intensity which is ~

7 fold higher than that upon direct excitation of Fl at 490 nm in the absence of the polymer, indicating signal amplification provided by the CCP. In contrast, the polymer sensitized fluorescent signal of the nonspecific ion treated sample is only amplified ~4 fold. The high signal amplification of using fluorescent CCP is important to reduce false-positive signal in the process of mercury(II) detection.

CONCLUSIONS

In summary, we have demonstrated a fluorescence turn-on assay for mercury(II) with an improved performance. This sensor system takes advantage of signal amplification of fluorescent CCP and NS-based detection in addition to the specific T-Hg^{2+}-T interaction. The use of CCP significantly enhances the detection fluorescence intensity and reduces false-positive signal. Moreover, the CCP also increases the signal ratio of Hg^{2+}/nonspecific ionic mixture and improves the detection selectivity. The developed sensing platform (CCP as the signal amplifier, silica NS as the solid support, and nucleic acid duplex as the recognition element) can be easily extended to detect other metal ions through the appropriate design of probe structure (either nature or artificial sequences) [2a], and even operated at room temperature [5b]. Therefore, it opens up more viable research opportunities in the field of chemical sensing.

REFERENCES

1. H. H. Harris, I. J. Pickering and G. N. George, *Science* **301**, 1203 (2003).
2. [2a] G. H. Clever, C. Kaul and T. Carell, *Angew. Chem. Int. Ed.* **46**, 6226 (2007); [2b] Y. Miyake, H. Togashi, M. Tashiro, H. Yamaguchi, S. Oda, M. Kudo, Y. Tanaka, Y. Kondo, R. Sawa, T. Fujimoto, T. Machinami and A. Ono, *J. Am. Chem. Soc.* **128**, 2172 (2006).
3. A. Ono and H. Togashi, *Angew. Chem. Int. Ed.* **43**, 4300 (2004).
4. X. Liu, Y. Tang, L. Wang, J. Zhang, S. Song, C. Fan and S. Wang, *Adv. Mater.*19, 1471 (2007).
5. [5a] J. S. Lee, M. S. Han and C. A. Mirkin. *Angew. Chem. Int. Ed.* **46**, 4093 (2007); [5b] X. Xue, F. Wang and X. Liu, *J. Am. Chem. Soc.* **130**, 3244 (2008).
6. J. Liu and Y. Lu, *Angew. Chem. Int. Ed.* **46**, 7587 (2007).
7. S. W. Thomas, G. D. Joly and T. M. Swager, *Chem. Rev.* **107**, 1339 (2007).
8. B. Liu and G. C. Bazan, *Chem. Mater.* **16**, 4467 (2004).
9. B. Liu and G. C. Bazan, *J. Am. Chem. Soc.* **126**, 1942 (2004).
10. Y. Wang and B. Liu, *Chem. Commun.* **34**, 3553 (2007).
11. [11a] B. S. Gaylord, A. J. Heeger and G. C. Bazan, *Proc. Natl. Acad. Sci. USA.* **99**, 10954 (2002); [11b] S. Wang, B. S. Gaylord and G. C. Bazan, *Adv. Mater.* **16**, 2127 (2004).
12. M. B. Aberem, A. Najari, H. A. Ho, J. F. Gravel, P. Nobert, D. Boudreau and M. Leclerc, *Adv. Mater.* **18**, 2703 (2006).
13. B. Liu and G. C. Bazan, *J. Am. Chem. Soc.* **128**, 1188 (2006).
14. Y. Wang and B. Liu, *Anal. Chem.* **79**, 7214 (2007).
15. We first investigate the sensing method by comparing Hg^{2+} incubation during hybridization and after hybridization. The results show that ~14.1% PL intensity was retained after thermal washing (45 °C) for mercury incubation during hybridization. In contrast, ~72.3% PL intensity was retained after thermal washing (45 °C) for mercury incubation after hybridization. Therefore, the sensing scheme of using mercury incubation after hybridization is selected for our Hg^{2+} assay.

Mater. Res. Soc. Symp. Proc. Vol. 1134 © 2009 Materials Research Society   1134-BB05-02

Novel Enzymatic Polymerization of Diazo Compounds: A New Group of Dyes

Ferdinando F. Bruno, Lauren E. Belton and Diane M. Steeves [1]

[1] U.S. Army Natick Soldier Research, Development & Engineering Center, Natick, MA 01760, U.S.A.

ABSTRACT

The homopolymerization of Sudan Orange G, Disperse Orange 3, Calcon, Mordant Yellow 12, Disperse Orange 13, and Bismarck Brown Y, was mediated by the enzyme horseradish peroxidase. Fourier transform infrared (FTIR) spectroscopy and matrix assisted time-of-flight mass spectrometry (MALDI-TOF) was utilized to confirm the polymeric structures. The enzymatic reaction produced novel electroluminescent polymers that exhibit a red or a blue shift, and enhanced non-linear optical behavior when compared to the corresponding monomers. The results of detailed optical analysis of the polymeric materials will be presented.

INTRODUCTION

In the past 20 years enzymatic polymerization has emerged as a technique for the formation of a variety of macromolecules [1-5]. The mild polymerization conditions, environmental compatibility and high selectivity of enzyme catalyzed polymerizations are clear advantages of this procedure compared to traditional polymerization procedures. The enzyme horseradish peroxidase is frequently implemented to polymerize complex macromolecules that can not be produced by classical chemical procedures. As an example, horseradish peroxidase has been utilized to catalyze the polymerization of phenol and aniline derivatives for the formation of conductive, conjugated polymers [5-6].

The synthesis and characterization of polymers in monophasic organic solvents with peroxidase in dioxane has been reported [6]. A wide range of substituted phenols and anilines were reactive under these conditions and formed homo- and co-polymers. High molecular weight polymers (400K) were synthesized that were stable at high temperatures and exhibited melting points in the 215-250°C range. The substitution pattern for chain or network growth was dependent on the monomers and their ring substituents [6-8]. The products are polymers with useful electronic and optical properties. However, the poor solubility and high branching or 3-dimensional network configuration of these polymers prevented easy processing into films or fibers [4-5]. An approach was sought to overcome the processing limitations while maintaining or enhancing the optical and electronic properties of the polymers. We report here a biocatalytic approach for the enzymatic polymerization in water-organic solvent of different dyes which accomplishes this goal.

EXPERIMENTAL DETAILS

The dyes Sudan Orange G (SOG), Disperse Orange 3 (DO3), Calcon (CAL), Mordant Yellow 12 (MY12), Disperse Orange 13 (DO13), and Bismarck Brown Y (BBY) were purchased from Aldrich Chemical Co. (Milwaukee, WI) and used as received. The chemical structures of these dyes are shown in figure 1. The monomers SOG, CAL and DO13 in 1-2 mg/ml concentration were solubilized in dimethylformamide (DMF) and 2 to 4 ml of this solution was added to 8 to 6 mL of buffered water at pH 7.5. The water contained 3 mg of horseradish peroxidase (HRP, EC 1.11.1.7 Type II, 150-200 units/mg), that was purchased from Sigma-Aldrich Chemical Co. (St. Louis, MO). The polymer formation of different dye monomers was initiated at room temperature by the injection of 2 ml of 1% hydrogen peroxide solution (Aldrich Chemical Co.) into the buffered water/DMF solution. Reactions were complete after 12 hours and the resultant precipitated polymers were dried and transferred onto various substrates for characterization. The same procedure was used for DO3, MY12, and BBY; however the pH for these reactions was 4.3.

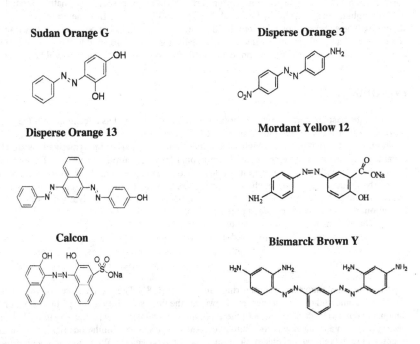

Figure 1. Chemical structures of dye monomers utilized in this study.

The polymers were soluble in DMF, and not in water, with the exception of polyBBY, which was not soluble in any solvent. Number average molecular weights (Mn) of the polymers were determined using matrix assisted time-of-flight (MALDI-TOF) mass spectrometry (MALDI LR, Waters, Milford, MA, USA, formerly Micromass, UK). The samples were dissolved in DMF and mixed with an equal concentration of the matrix 2,5 dihydroxybenzoic acid (Aldrich, USA) for the analysis. Spectral characterization of monomers and polymers were performed with a Perkin-Elmer Lambda-9 UV-Vis-Near IR spectrophotometer (Norwalk, CT). Fourier transformed infrared (FTIR) characterization of monomers/polymers deposited on NaCl plates was performed on a Perkin Elmer FTIR 1720X. The nonlinear optical behavior of the dyes was investigated using a 113 mW Ar/Kr laser at 530 nanometers. All samples were dissolved in DMF at concentrations resulting in ~50%T at 530 nm. Figure 2 shows the experimental configuration consisting of an Ar/Kr continuous wave laser as the source, an electrical shutter, polarizer, and photodetector. A Lecroy digital oscilloscope was used for the data acquisition.

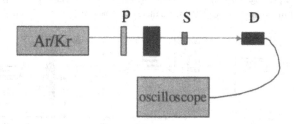

Figure 2. Experimental configuration used to measure nonlinear properties. The light source is an Ar/Kr laser; P, polarizer; S, shutter; sample; D, photodetector, and oscilloscope.

DISCUSSION

The number average molecular weights of the polymerized dyes were obtained using Maldi-TOF techniques. The spectrum for the polymerized form of DO13 is shown in Figure 3.

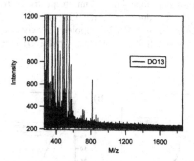

Figure 3. MALDI-TOF spectrum of polymerized sample of DO13 in DMF.

Table 1 lists the Mn values for the various polymerized samples. The Mn values range from 789.5 for MY12 to 1037.3 for SOG. It is obvious from these values that the polymerization process resulted in short-chain polymers (oligomers) rather than long-chain polymers.

Sample	Molecular Weight (M$_n$)
DO13	881.2
CAL	909.1
MY12	789.5
DO3	831.6
SOG	1037.3

Table 1. Number average molecular weights (Mn) for the polymerized dyes.

The UV-Vis spectrum of the product obtained by enzymatic polymerization of SOG, DO3 and DO13 in the presence of HRP were measured and are shown in figure 4.

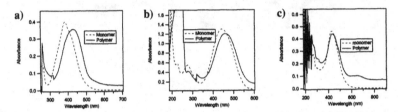

Figure 4. UV-Vis spectra of a) polySOG and of SOG monomer, b) polyDO3 and DO3 monomer and c) polyDO13 and DO13 monomer.

In the region from 400 to 500 nm, a red shift of 48 nm, 21 nm and 8 nm in the absorption is observed for polySOG, polyDO3 and poly DO13, respectively. This indicates an increase in the degree of conjugation in the final products. This absorption shift was observed previously [7] with the enzymatic polymerization of phenol and of aniline. However, in the polymerization of CAL and MY12 a blue shift of 261 nm and of 112 nm, respectively, was observed, as shown in figure 5. This blue shift is probably due to the three dimensional configuration of the polymer. It is possible that for polyCAL and polyMY12 the molecules exist in a trans configuration. However, for polymers such as polySOG, polyDO3 and polyDO13 the polymers likely exist in a cis type configuration [9].

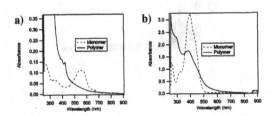

Figure 5. UV-Vis spectra of a) polyCAL and CAL monon..., and b) polyMY12 and MY12 monomer.

Figure 6 shows the Fourier transform infrared (FTIR) spectra of the polyMY12 and of the monomer MY12. Two important absorptions are observed in the spectrum collected for the polymer. The first is a significant broadening and shifting at lower frequency in the NH stretch region (3000-3500 cm^{-1}) of the polymer with respect to the monomer. This is explained by stronger hydrogen bonding in the polymer and by the more rigid structure created after the polymerization [7]. The second absorption peak of interest is in the region 1550-1690 cm^{-1} that is attributed to conjugation in the main chain. These FTIR data corroborate the findings of the UV-Vis spectra.

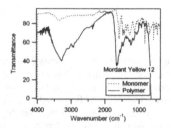

Figure 6. FTIR collected for MY12, and for polyMY12. Similar results were observed for polySOG, polyDO3, polyDO13, and polyCAL.

The non-linear optical behavior of the monomeric and polymerized dyes was investigated using a 113 mW Ar/Kr continuous wave laser at a wavelength of 530 nm. The dyes were dissolved in DMF and the response was measured in units of voltage as a function of time. The voltage can be directly correlated to transmission. The NLO response of the monomeric dyes, from fastest to slowest, is as follows: SOG > DO3 ≈ DO13 > MY12 > CAL. The response of the best (SOG) and worst (CAL) monomeric dyes is shown in figure 7a.

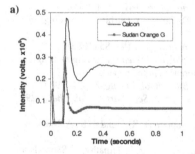 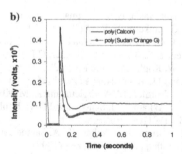

Figure 7. NLO response of a) the best (SOG) and worst (CAL) monomeric dyes; and b) the most improved (polyCAL) and least improved (polySOG) polymerized dyes.

The polymerized dyes exhibited nearly the same order in terms of response time: SOG ≈ DO13 > DO3 > MY12 > CAL. The NLO response of the polymerized dyes was generally enhanced in comparison to the monomeric dyes. The greatest enhancement was observed for CAL, which had the worst performance as a monomer. The response of the most improved (polyCAL) and least improved (polySOG) polymerized dyes is shown in figure 7b. The improved NLO response of the polymerized dyes may be due to a decrease in the trans-cis isomerization rate realized by a cooperative molecular dynamics scheme along the polymer chain that is not possible with the discrete monomers in solution. Further investigation is needed to fully elucidate the nature of this behavior.

CONCLUSIONS

The biocatalytic approach for polymeric dye synthesis described is a free radical polymerization process. A wide range of monomers will react under these conditions to provide a diversity of potential polymer products for systematic studies of the effect of monomer constituents on mechanical, thermal, electronic and linear and non linear optical properties. The process described here represents a general technique for the assembly and polymerization of optically active polydyes. Processing limitations observed with intractable polymers synthesized in bulk are overcome with this technique allowing the preparation of thin films by spin coating techniques.

ACKNOWLEDGEMENTS

The authors gratefully acknowledge the Center for Advanced Materials at the University of Massachusetts Lowell for assistance with the Maldi-TOF analysis.

REFERENCES

1. T. Kobayashi, H. Yoneyama, H. Tanura, *J. Electronal Chem.* **161**, 419 (1984).
2. V.G. Kulkarni, W.R. Matthew, J.C. Cambell, C.J. Dinkine, P.J. Durbin, *Proceedings of the Society of Plastics Engineering* (1991) p. 665.
3. Genies, E.M., Hany, P., Sentier, C.J. *J Appl. Electrochem. Soc.* **18**, 285 (1982).
4. P. W. Kopf, "Phenolic Resins," in *Encyclopedia of Polymer Science and Engineering*, Volume 11, Wiley, New York, (1985) p. 45 (and references in the paper).
5. J. S. Dordick, M. A Marletta and A. M. Klibanov, *Biotechnol. Bioeng.* **30**, 31 (1987).
6. J. A. Akkara, K. J. Senecal, and D. L. Kaplan, *Jour. of Pol. Sci.: Part A: Pol. Chem.*, **29**, 1561, (1991).
7. F. F. Bruno, J. A. Akkara, L. A. Samuelson, D. L. Kaplan, B. K. Mandal, K. A. Marx, S. K. Tripathy, *Langmuir* **11**, 889 (1995).
8. F. F. Bruno, R. Nagarajan, J. S. Sidhartha, K. Yang, J. Kumar, S. Tripathy, L. Samuelson, *Mat. Res. Soc. Symp. Proc.* **600**, 255 (1999).
9. P. Wu , S. Balasubramanian, W. Liu, J. Kumar, L. Samuelson, S. K. Tripathy, *Jour. Macr. Molec. Scien. Part A*, **38 (12)**, 1463 (2001).

Mater. Res. Soc. Symp. Proc. Vol. 1134 © 2009 Materials Research Society   1134-BB05-04

Tunable Quantum Confinement Effect on Non-Volatile Thin Film Polymer Memory Device

Augustin J. Hong[1], Kang L. Wang[1], Wei Lek Kwan[2], Yang Yang[2], Dayanara Parra[3] and Sarah Tolbert[3]

[1]Department of Electrical Engineering, [2]Department of Materials Science and Engineering, [3]Department of Chemistry, University of California Los Angeles, Los Angeles, California 90095, USA

ABSTRACT

Several dodecanethiol capped Au nanoparticles (Au-DT) less than 5 nm in size were synthesized for the fabrication of a non-volatile polymer memory device (PMD). Size characterizations of Au-DTs were confirmed by XRD and ^{1}H NMR. The PMD consisted of a polystyrene thin film containing Au-DT and 8-Hydroquinoline (8-HQ) sandwiched between two aluminum electrodes. The device exhibited desired "write", "read" and "erase" steps of a memory cycle. Quantum confinement effects were apparent because electronic properties of the device depended on the nanoparticle size. An energy band diagram for the quantized charging and a novel conducting mechanism were suggested.

INTRODUCTION

As the state-of-the-art memory technology continues to increase its density, the metal-oxide-semiconductor field effect transistor (MOSFET) is approaching its physical limit. Among various memory structures, the cross bar type memory readily becomes the type of most interest because it is the simplest memory type that can be scaled down easily beyond current MOS technology. In order for this cross bar type memory to be realized, new functional materials that can act as logic "0" or logic "1" with bias must be synthesized and placed between two electrodes.

Toward this end, various approaches using organic molecules, inorganic thin film and inorganic-organic hybrid thin film structures have been demonstrated. To date, Green et. al. demonstrated the promising volatile memory application beyond MOS technology using an organic molecule, the bistable [2] rotaxanes.[1] For non-volatile memory applications, Ouyang et. al. showed the new cross bar type non-volatile memory device using inorganic-organic hybrid thin film structure, which can be easily fabricated, cost effective and showed high speed and low power consumption compared to current flash memory technology.[2] Also, Jeong et. al. suggested a simple metal-inorganic-metal thin film cross bar structure that can be used for non-volatile memory application.[3] However, the dual mode behavior (bipolar and unipolar) of this inorganic thin film according to its current level makes it less attractive toward the memory applications.

In this study, we reproduced the non-volatility of polymer based thin film memory device and demonstrate the tunability of its operation point which can be explained by the quantum confinement effect.

EXPERIMENT

The Au-DT nanoparticles with different size distributions were synthesized by varying the mole ratio of dodecanethiol:Au. To a vigorously stirred solution of 0.5 g tetraoctylammonium bromide (TOAB, [CH$_3$(CH$_2$)$_7$]$_4$NBr, 2.5 equivalent (eq.)) in 25 mL 1,2-dichlorobenzene, a solution of 0.1 g hydrogen tetrachloroaurate(III) trihydrate (HAuCl$_4$•3H$_2$O, 1 eq.) in 8 mL deionized water was added. The dark orange hydrogen tetrachloroaurate(III) trihydrate aqueous solution cleared within 5 minutes and the toluene phase became orange-brown as the AuCl$_4^-$ was transferred into it. Once the organic phase was isolated, the desired amount of dodecanethiol was added (61, μL, 1 eq.; 122, μL, 2 eq.; 12, μL, 0.2 eq.; 6.1, μL, 0.1 eq.). The resulting solution was stirred for 10 min at room temperature. The solvent was removed by rotary evaporation and ethanol was added to the black solid. The solution was sonicated for 30 minutes and then transferred to an Erlenmeyer flask. The black solid settled at the bottom of the flask after an hour and the solvent was decanted. The black products were dried under vacuum for 24 hours. In order for the size estimation using XRD, some black products were dissolved in hexanes to yield a white precipitate which was then filtered through a 20 nm pore filter paper. The filtrate was collected and evaporated to dryness. XRD spectra were obtained using the Panalytical Expert Pro. ^{1}H NMR spectra were obtained in deuterated benzene (C$_6$D$_6$) on a Bruker AV300 spectrometer. NMR spectra are reported in units of ppm relative to C$_6$D$_6$ (7.16 ppm).

After the size characterizations, device fabrication was started on a high resistivity (21 Ωcm ~ 28 Ωcm) 4 inch p type silicon wafer with 200nm SiO$_2$ layer deposited on top using plasma enhanced chemical vapor deposition (PECVD). Al bottom electrode was deposited using CHA e-beam evaporator under less than 5×10^{-7} torr. A 50nm active layer of 1,2 dichlororbenzenic solution with 0.4wt% 1-dodecanethiol capped Au nanoparticles (Au-DT), 0.4wt% 8-hydroquinoline (8HQ), and 1.2wt% polystyrene (PS) was spin coated. Finally, Al top electrodes were deposited perpendicular to the bottom electrodes. Al electrodes were 200nm thick and 2mm wide, giving a device area of 4mm^2. The electrical properties of the polymer-Au DT hybrid memory devices were characterized with a Keithley semiconductor characterization system 4200.

RESULTS AND DISCUSSION

Materials Characterizations

X-ray diffraction pattern (XRD) was used to estimate the averaged nanoparticle sizes. The data in Fig. 1 confirms phase purity of our particles based on the Au (111) crystalline peak at 2θ= 39. Peak broadening is attributed to the small particle size and the extraneous peaks correspond to the phase transfer agent TOAB which was not fully removed prior to filtration. From the Scherrer equation (eq.1), we can estimate the diameter (D) for the 1:10 dodecanethiol:Au nanoparticles which was 3.7 nm, and the diameter for 1:1 sample which was 1.8 nm.

$$L = \frac{0.9\lambda_{x-ray}}{\Delta(2\theta)\cos\theta} \quad (1)$$

where L=0.75D for sphere, λ=0.154056 nm (Cu, Kα1, 2θ=39 degree, and Δ(2θ)=full width at half maximum in radian.

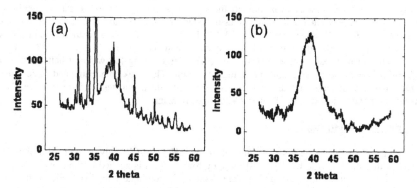

Figure 1. X-ray diffraction patterns of gold nanoparticles (a) dodecanethiol:Au = 1:10, D = 3.7 nm; (b) dodecanethiol:Au = 1:1, D = 1.8 nm.

These values are in consent with the size calculation by Hostetler et. al.[4] No XRD data could be obtained for the 1:5 or 2:1 samples because they could not form powders.

To further investigate the size increase in all four samples that were synthesized, NMR was employed. NMR can be a useful tool in elucidating size differences due to anisotropic effects that arise with size increase. It is known that for large molecules such as polymers and proteins NMR peaks are broad and undefined due to decrease in rotation and Brownian motion in solution. This decrease in motion causes inhomogeneously induced magnetization in transverse plane perpendicular to the magnetic field. This causes a decrease in spin-spin relaxation (T_2 relaxation). T_2 relaxation is the value that determines the peak width, and so the shorter the relaxation time the broader the peaks become. Therefore the ^{1}H NMR peaks should broaden with an increase in the size of the particles.

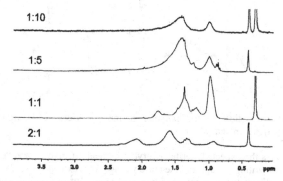

Figure 2. ^{1}H NMR of Au-DT nanoparticles with several different dodecanethiol:Au ratios.

This is in fact what is observed. Figure 2 shows the peaks at 2.1 ppm, corresponding to the methylenes (CH_2) in closest proximity to the Au nanoparticle, in the 2:1 sample begin to coalesce with the peaks at 1.6 and 1.3 ppm as the dodecanethiol:Au ratio decreases. These peaks continue to coalesce as ratio decreases from 1:1 to 1:10. Also, spin-spin relaxation time (T_2) was calculated to be shorter (0.15sec., 0.13sec., 0.09sec. and 0.08sec. for 2:1, 1:1, 1:5 and 1:10 dodecanethiol:Au ratios) as nanoparticle size became larger using full width at half maximum (FWHM) of the methyl (CH_3) signal peak around 0.9ppm. This peak broadening and T_2 decrease suggest the size increase and thus the control the synthetic parameters had on the size. The typical number of scans was 8 in Bruker AV300 spectrometer.

Electrical Characterizations

Figure 3a shows the memory function of the 8HQ+Au-DT+polymer thin film memory device with a thiol to gold ratio of 1:10. When a forward bias is applied to the device, the I-V curve shows an initial low conducting state. Voltage is continually applied until there is an abrupt transition to high conducting state at around 2.5V (curve A). During the subsequent voltage scan (curve B), device shows good stability in high conducting state. This means that once logic "1" is programmed at more than 2.5V, we can read this logic "1" at the reading voltage of 1V. To erase the information stored in the device, in the reverse bias is applied. Curve C depicts this, the device resets to the initial low conducting state as -1.2 V is supplied. After 10 cycles of Write-Read-Erase operation, transition points remain same. On/Off ratio at the ambient environment was close to 10^3. The I-V curve of device with a thiol to gold ration of 1:1 is shown in Figure 3b.

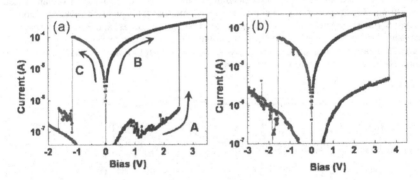

Figure 3. Memory function of the polymer memory devices made from two different nanoparticle sizes (a) 3.7nm and (b) 1.8nm.

It also depicts the expected "write" (Curve A), "read" (Curve B) and "erase" (Curve C) cycle. However, the writing voltage and the erase voltage were extended to 3.7V and -1.7V each. Also the On/Off ratio at the ambient condition was about 10^2. To explain the differences in transition points and the On/off ratio, an energy band diagram is suggested as figure 4a. In positive bias scheme, device needs energy that overcomes both tunneling barrier through alkanethiol and energy difference between HOMO level and Fermi energy level of the Au nanoparticle. As Au nanoparticles become smaller, the Fermi energy level of Au increases due to the quantum confinement effect. The electrons from the HOMO of 8HQ (6.1 eV) [5] therefore require more energy to be transferred to Au nanoparticles. This quantization of the charging energy causes different transition point with different nanoparticle sizes.

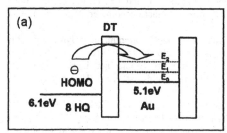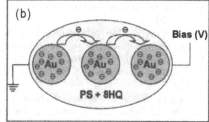

Figure 4. (a) Energy band diagram for the quantized electron charging of Au-DT nanoparticles (b) Proposed electron hopping mechanism between charged Au particles creating high conducting state of the polymer memory

The asymmetrical behavior in transition point when device is negatively biased is due to the energy level difference between Al work function (4.1 eV) and 8HQ HOMO level (6.1 eV) which makes an energy barrier for holes when holes transfer from Al electrode to 8HQ. This barrier disappears when the holes transfer in the reverse direction. It is not yet fully understood how the high conducting states was achieved once the nanoparticles become charged. We speculate that once the Au nanoparticles are charged by electrons, the high conducting state can be achieved through electron hopping between charged gold nanoparticles as depicted in figure 4b because the charged nanoparticles has a shallower energy barrier than initially uncharged nanoparticles therefore higher chance for the electron hopping to occur.

CONCLUSIONS

In summary, this experiment depicts the importance of size of nanoparticles on electronic properties of polymer memory devices. Different size distributions of Au nanoparticle were confirmed using XRD and NMR. Quantization of charging energy because of the quantum confinement effect explains different transition points of polymer memory devices with different nanoparticle sizes. Since this quantum confinement effect is tunable by the simple synthetic method that can be easily controlled, this work may provide a simple and reliable approach to create polymer memory devices with desired electrochemical properties.

Authors would like to thank the NSF IGERT: Materials Creation Training Program (MCTP) - DGE -11443 and the California NanoSystems Institute (CNSI) for their financial support.

REFERENCES

1. J.E. Green, J.W. Choi, A. Boukai, Y. Bunimovich, E. Johnston-Halprin, E. DeIonno, Y. Luo, B.A. Sheriff, K. Xu, Y.S. Shin, H.-R. Tseng, J.F. Stoddart, and J.R. Heath: A 160-kilobit molecular electronic memory patterned at 10^{11} bits per square centimeter. *Nature 445*, 414-417 (2007).
2. J. Ouyang, C. Chu, C. R. Szmanda, L. Ma, and Y. Yang: Programmable polymer thin film and non-volatile memory device. *Nature Mater.* Vol. 3, No.12, 918-922 (2004).
3. D. S. Jeong, H. Schroeder, and R. Waser: Coexistence of Bipolar and Unipolar Resistive Switching Behaviors in a Pt/TiO$_2$/Pt Stack. Electro. Solid-State Lett. **10** , 8, G51-G53 (2007).
4. M. J. Hostetler, J. E. Wingate, C. J. Zhong , J. E. Harris, R. W. Vachet, M. R. Clark, J. D. Londono, S. J. Green, J. J. Stokes, G. D. Wignall, G. L. Glish, M. D. Porter, N. D. Evans, R. W. Murray: Alkanethiolate Gold Cluster Molecules with Core Diameters from 1.5 to 5.2 nm: Core and Monolayer Properties as a Function of Core Size. *Langmuir* 1998, *14,* 17-30.
5. J. Ouyang, C. Chu, R. J. Tseng, A. Prakash, and Y. Yang: Organic Memory Device Fabricated Through Solution Processing. Proc. IEEE, VOL. 93, NO. 7, 1287-1296 (2005)

Mater. Res. Soc. Symp. Proc. Vol. 1134 © 2009 Materials Research Society   1134-BB05-07

Novel pH-Thermosensitive Gel Adsorbents for Phosphoric Acid

T. Gotoh, T. Arase and S. Sakohara
Department of Chemical Engineering, Graduate School of Engineering, Hiroshima
University, Higashi-Hiroshima 739-8527, Japan

ABSTRACT

A novel pH-thermosensitive gel adsorbent for phosphoric acid has been developed. The adsorbent was prepared by the copolymerization of a thermosensitive polymer and a polymer with tertiary amino groups. Phosphoric acid was adsorbed and desorbed repeatedly with a change in pH and temperature. The phosphoric acid was adsorbed onto the gel at low pH and low temperature, and it was desorbed from the gel at high pH and high temperature. The amount of adsorption increased when a pH value was decreased, because the tertiary amino groups were not completely ionized at high pH. The amount of adsorption decreased with an increase in the initial pH of the solution. The amount of desorption of the phosphoric acid further increased with an increase in temperature because the shrinkage of the gel at higher temperature enhanced the desorption of phosphoric acid from the gel. The application of a novel pH/temperature-sensitive gel to enhance the adsorbent qualities of phosphoric acids in water is proposed.

INTRODUCTION

The eutrophication of closed water such as lakes, ponds or bays by an influx of phosphorous has become a serious problem. On the other hand, phosphorous is one of the essential elements for life, and should be recovered and reused, as it is a valuable, limited resource. Various methods for the removal or recovery of phosphate from wastewater have been proposed. However, the purity of the recovered phosphorous was not high enough to permit its reuse. Thus, the purification of phosphate is needed. Ion exchange is one of the methods for purifying the recovered phosphorous, but the regeneration of ion-exchange resins is a costly process that creates new waste. A novel pH-thermosensitive gel adsorbent for phosphoric acid has been developed and its adsorption/desorption properties are investigated in this research.

CONCEPT

Dissociation of Phosphoric Acid

Phosphoric acid varies its state of dissociation with a change in pH. The state of dissociation affects adsorption of the phosphoric acids. Phosphoric acid dissociates in three steps, as expressed by equation (1), (2) and (3).

$$H_3PO_4 \leftrightarrow H_2PO_4^- + H^+ \quad (pK_{a1} = 2.15) \tag{1}$$

$$H_2PO_4^- \leftrightarrow HPO_4^{2-} + H^+ \quad (pK_{a2} = 7.20) \tag{2}$$

$$HPO_4^{2-} \leftrightarrow PO_4^{3-} + H^+ \quad (pK_{a3} = 12.35) \tag{3}$$

The prepared gel adsorbent that has tertiary amino groups in the thermo-sensitive gel network is composed by N,N-dimethylaminopropylacrylamide (DMAPAA) and N-isopropylacrylamide (NIPA). The adsorbent has a lower critical solution temperature (LCST) and becomes hydrophobic above the LCST. The tertiary amino groups of DMAPAA were ionized in water and shows a cationic state as expressed by equation (4).

$$CH_2CHCONH(CH_2)_3N(CH_3)_2 + H_2O \leftrightarrow CH_2CHCONH(CH_2)_3N^+H(CH_3)_2 + OH^- \quad (4)$$

The ionization of the tertiary amino group is used for the adsorption of phosphoric acid.

Mechanism of Adsorption / Desorption

The valence of phosphoric acid increases with an increase in pH. The concept of adsorption/desorption by the thermo /pH-sensitive gel is illustrated in Figure 1. Univalent phosphoric acid is adsorbed to ionized tertiary amino groups at low pH. The univalent phosphoric acid then turns to a divalent one when the pH is raised. The divalent phosphoric acid is desorbed because the degree of ionization of the tertiary amino groups of the gel is decreased at high pH. The phosphoric acid is further desorbed by shrinkage of the gel when this gel adsorbent is heated above the LCST. This is because the ionization of the amino groups in the gel network was suppressed by a decrease in the water content of the gel above the LCST. The phosphorous ions, which are free from the adsorption site, were expelled from the gel by its shrinkage.

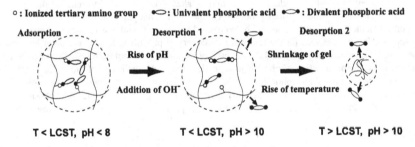

Figure 1 Mechanism of adsorption and desorption

EXPERIMENT

The N-isopropylacrylamide (NIPA) and N,N'-dimethylaminopropylacrylamide (DMAPAA) was kindly supplied by Kohjin Co. (Japan). The NIPA was purified by recrystallization from n-hexane and DMAPAA was purified by reduced distillation. All other chemicals were purchased from Sigma-Aldrich Co. (USA). They were reagent grade and used without further purification. The preparation condition of the gel is shown in Table 1. The copolymerization of NIPA and DMAPAA was conducted by radical polymerization in Teflon tubes of 6mm inner diameter at 80 °C, higher than the LCST for a NIPA polymer, to make the gel porous for quick swelling and shrinking [1]. After completion of the reaction, cylindrical gels were washed with methanol by Soxhlet extraction to remove unreacted chemicals. The cylindrical gels were then sliced into pieces of 1.5 mm thickness and dried at room temperature.

The adsorption/desorption experiment was conducted as follows. The phosphoric acid solutions of different valences (potassium phosphate monobasic, potassium phosphate dibasic and potassium phosphate tribasic) were prepared as the adsorbate solution. The desired amount of the dried gels was then added to the solution. The solutions with the gel were shaken in a water bath at a desired temperature for 24 hours. Then, the gels which adsorbed phosphoric acid were added to a sodium hydroxide solution with a different pH to desorb the phosphoric acid at 20 °C. The temperature of the solution was then raised to 50°C for additional desorption of phosphoric acid. The amount of phosphate adsorbed on the gel was determined from the decrease in the concentration of phosphate in the solution. The concentration of the phosphate was determined by ion chromatograph.

Table I. Preparation condition of the gel

	Material	[mol/m^3]
Monomer	N-isopropylacrylamide	1900
Co-monomer	N,N-dimethylaminopropylacrylamide	100
Cross-linker	N,N'-methylenebisacrylamide	100
Accelerator	N,N,N',N'-tetraethylmethylenediamine	10
Initiator	Ammonium peroxodisulfate	0.5

DISCUSSION

Adsorption of phosphoric acids on the gel

Figure 2 shows the adsorption isotherm of phosphoric acid with different valences adsorbed in the gels. The amount of adsorption increased when the initial pH value was decreased, because the tertiary amino groups were not completely ionized at the high pH. Therefore, more phosphoric acid was adsorbed at low pH than at high pH. Figure 2 also shows that the amount of adsorption does not increase proportionally to its valence. For example, the ratio of the amount of adsorption of divalent phosphoric acid to that of univalent is less than half. This is because the pH of the divalent phosphoric acid solution is higher than that of the univalent one and ionization of amino groups in the polymer gel networks is suppressed in the solution at high pH.

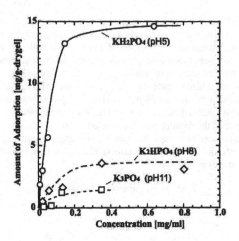

Figure 2 The adsorption isotherm of phosphoric acid of different valences

Figure 3 shows the effect of the valences of the phosphoric acid on the swelling diameter of the gel. The gel swells larger in univalent phosphoric acid than in divalent phosphoric acid. This suggests that the divalent phosphoric ion that was adsorbed in the gel behaved so as to suppress the swelling of the gel like a cross-linker. The results expressed in figures 2 and 3 suggest the propriety of the concept shown in figure 1.

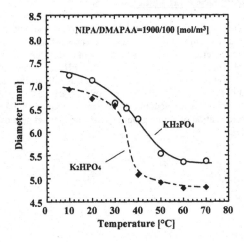

Figure 3 The effect of valences of phosphoric acids on the swelling property of the gel

Desorption of phosphoric acids from the gel

Figure 4 shows the effects of the initial pH value of the sodium hydroxide solution on the amount of desorption of phosphoric acids. The amount of desorption increased with increasing pH of the solution. An increase in pH suppresses the ionization of amino groups in the gel network. The degree of dissociation of phosphoric acid increased with an increase in pH and those reactions encouraged the phosphate acid to desorb from the gel. The amount of desorption of the phosphate acid further increased with an increase in temperature, because the shrinkage of the gel at higher temperature enhanced the desorption of phosphoric acid from the gel as expressed in figure 1.

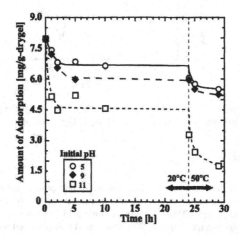

Figure 4 Effect of pH and temperature on the desorption of phosphoric acids

Figure 5 shows the change in the amount of adsorption of phosphoric acid on the gel as a result of the pH and temperature swing. The phosphoric acid repeatedly adsorbed and desorbed in the process. The amount of adsorption at 20 °C decreased with an increase in repeating times. Because the desorbing solution remained in the gel during the desorbing procedure suppress the ionization of amino groups in the gel at the adsorbing process, and the amount of adsorption of phosphoric acid decreased.

107

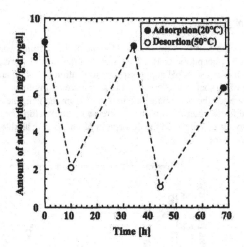

Figure 5 Adsorption/desorption of phosphoric acids in response to the pH/temperature swing. Concentration of phosphate at adsorption was 5 mM. The value of pH at desorption was 11.

CONCLUSIONS

A novel pH-thermosensitive gel adsorbent for phosphoric acid has been developed. The gel was prepared by the copolymerization of a thermosensitive polymer (NIPA) and a polymer with tertiary amino groups (DMAPAA). Phosphoric acid was adsorbed and desorbed repeatedly with a change in pH and temperature. Phosphoric acid was adsorbed on the gel at low pH and low temperature, while it desorbed from the gel at high pH and high temperature. The amount of adsorption increased when the initial pH value was decreased, because the tertiary amino groups were not completely ionized at high pH. The amount of adsorption decreased when the initial pH of the desorbing solution was increased. The amount of desorption of the phosphate acid further increased with an increase of the temperature because the shrinkage of the gel at higher temperature enhanced the desorption of phosphoric acid from the gel. These results suggest that the application of this novel pH/temperature sensitive gel could enhance the absorbency of phosphoric acids in water.

ACKNOWLEDGMENTS

This research was partly supported by a Grant-in-Aid for Scientific Research (18560723, 20560704) from the Japan Society for the Promotion of Science and Research for Promoting Technological Seeds (10-073) from the Japan Science and Technology Agency.

REFERENCES

1. T. Gotoh, Y. Nakatani and S. Sakohara, *J. Appl. Polym. Sci.*, **69,** 895 (1998)

Mater. Res. Soc. Symp. Proc. Vol. 1134 © 2009 Materials Research Society   1134-BB05-10

Synthesis of Poly(Methylmethacrylate) Latex With Enhanced Rigidity Through Surfactant Control

Alex H.F. Wu[a], K.L. Cho[b], Irving I. Liaw[b], Hua Zhang[b] and Robert N. Lamb[b,c]

[a] School of Chemistry, The University of New South Wales, Sydney, New South Wales, Australia
[b] School of Chemistry, The University of Melbourne, Melbourne, Victoria, Australia
[c] Australian Synchrotron Company Ltd., Clayton, Victoria 3168, Australia

ABSTRACT

Water-based polymer latexes have attracted much attention since their invention in the early 1950s. Its advantages for both general health and the environment were recognized as they emit far less volatile organic compounds (VOCs) than their solvent-based counterparts.

The performance of latex paints and coatings is directly proportional to the ease of particle deformation. This is the main driving force for the paints and coatings industry to focus its research efforts towards understanding its mechanism.

In contrast, little has been published with respect to enhancing latex's resistance against deformation despite such needs in applications such as templating porous ceramics for catalysis and biomaterial engineering. Specifically, the latex's resistance to deformation is crucial to retain a network of uniform pores for applications relating to enzyme immobilization and materials reinforcement.

The current study reports increased heat-resistance observed in latexes when synthesized using a rigid surfactant, dimethyl siloxane – ethylene oxide block copolymer (PDMS-PEO). The film formation process for this latex was deduced using atomic force microscopy and subsequent roughness analysis. A comparative study using a non-ionic long-chain hydrocarbon surfactant, morpholine oleate, was also conducted.

INTRODUCTION

Polymer latexes have attracted much attention due to its potential application in drug delivery [1], enzyme immobilization [2] and materials templating [3]. Since its invention in the early 1950s [4], water-based latex paints and coatings were recognised as advantageous for both the environment and health as it emits far less volatile organic compounds (VOCs) than their solvent-based counterpart [5].

The rigidity and heat-resistance of latexes can be measured as a function of film formation temperature and is defined as the minimum temperature in which a latex film begins to collapse into a smooth film [5]. Film formation and heat-resistance of a polymer is primarily governed by the degree of cross-linking within its primary structure. To date, little has been published regarding the effect of surfactant on polymer latex rigidity and heat-resistance.

The current study reports the use of a polymerisation technique using a non-ionic siloxane surfactant, dimethyl siloxane – ethylene oxide block copolymer (PDMS-PEO) to produce monodisperse poly(methylmethacrylate) latex of enhanced rigidity and heat-resistance. A comparative study using a non-ionic long-chain hydrocarbon surfactant, morpholine oleate, was also conducted.

EXPERIMENT

All chemical reagents were used as received unless specified. Methyl Methacrylate (MMA) (99%, Aldrich), Milli-Q water (0.05µS/cm, Millipore [6]), Di (trimethylolpropane) tetracrylate (Aldrich), Dimethyl Siloxane – Ethylene Oxide Block copolymer (PDMS-PEO), 25% non-siloxane (400 cst, Gelest Inc, USA), Potassium persulfate (KPS) (>99%, Aldrich), Morpholine (>99%, BDH chemicals Ltd), Oleic Acid (Acid Chem International), Polyvinylpyrrollidone (Mw 30,000, Aldrich).

Synthesis of PMMA Nanoparticles

Morpholine Oleate-based PMMA emulsion polymerization was carried out in a reaction vessel charged with 100g water, 1.2g morpholine, 3.8g oleic acid, 10g MMA, 1.0g Di(trimethlolpropane) tetraacrylate and 0.1g KPS. The mixture was vigorously mixed with mechanical stirring by a hand-held mixer (Braun Multiquick 300 Watts) for 5 minutes. It was then purged with Argon for 10 minutes under constant stirring with a magnetic stirrer set at 700rpm. The mixture was then heated to 70-75°C for 3 hours with constant stirring. The same procedures were repeated for synthesizing PDMS-PEO based PMMA using 100g water, 5g PDMS-PEO block copolymer, 10g MMA, 1.0g Di(trimethylolpropane) tetraacrylate and 0.1g KPS. The ready emulsion was allowed to cool and filtered through cotton wool to remove coagulated PMMA.

Characterisation

In-Situ Particle Size Analysis
The resultant PMMA particle size and its polydispersity index were determined by Zeta Plus Particle Analyser (Brookhaven Instruments Corporation). Measurements were conducted at 25°C using a Helium/ Neon source to produce a wavelength of 660nm. Samples were placed in polystyrene cells with aqueous parameters (Viscosity: 0.89cP, Refractive index: 1.33). Five measurements were taken with the duration of 1 minute per measurement to obtain an average diameter.

Dried Particle Size Analysis
The particle size and shape of PMMA particles were observed with a Hitachi S-900 field emission scanning electron microscope (FESEM). Several drops of the latex suspension were placed on a glass slide and dried at 25°C and 110°C for 10 minutes. The dried PMMA particles were coated with a 10nm layer of chromium prior to imaging. The mean diameter was determined from the average of 100 latex particles.

Atomic force microscope images were captured using a Dimension-3000 AFM (Digital Instruments). Samples were analysed by tapping mode using an Otespa tip with a typical scan area of 5×5 μm².

DISCUSSION

At the completion of the polymerization, the resulting latex suspension appears milk-white and remained stable with no visible phase separation over several days. PMMA particles began to settle after 1 week and were readily re-dispersed with a small amount of agitation.

Surfactant system	Average Particle Size /nm	Polydispersity Index	Glass Transition Temperature (T_g) /°C
Morpholine Oleate (MO)	152	0.750	108
PDMS-PEO block copolymer	316	0.022	109

Table 1. Structural properties of PMMA particles synthesized using different surfactants

As summarized in Table 1, MO-based PMMA particles were observed to be significantly smaller and less monodisperse than PDMS-PEO block copolymer-based particles.

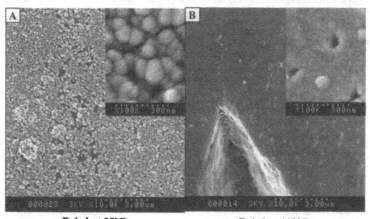

Dried at 25°C   Dried at 110°C

Figure 1. SEM images of MO based PMMA emulsion dried at (A) 25°C and (B) 110°C. Both images A and B were taken at 10, 000x magnification and their inserts were taken at 100, 000x magnification

Figure 1 shows SEM micrographs of the synthesized MO based PMMA particles dried at 25°C and 110°C. As can be seen from Figure 1a, the high polydispersity measured in-situ was confirmed as extensive physical deformation of MO-based PMMA particles were evident at 25°C. When dried at 110°C, the extent of the deformation was amplified to form a smooth film as shown in Figure 1b.

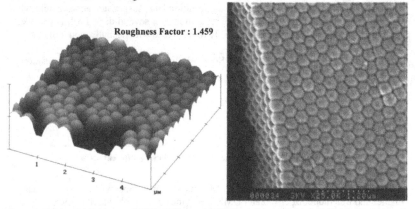

Roughness Factor : 1.459

Figure 2. SEM and AFM images PDMS-PEO based PMMA emulsion dried at 110 °C, showing ordered layers of close packed spheres

In contrast, shown in Figure 2, monodisperse PMMA particles were synthesized using PDMS-PEO as a surfactant with no visible deformation after drying at 110°C. Using roughness factor (R_f) as a measure of structural integrity of the drying PMMA particles, the correlation between roughness factor and drying temperature is shown in Figure 3.

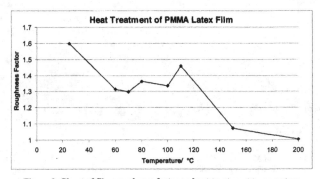

Figure 3. Chart of film roughness factor vs heat treatment temperature

The above data indicates that the PDMS-PEO based PMMA particles underwent significant deformation when dried above 110°C as the roughness factor for the latex film decreased dramatically. This decrease in roughness signifies the onset of film formation where polymer chain segments diffuse between particles, resulting in the observed flattening of the latex film.

The observed difference in particle size, polydispersity and film formation temperature between the two polymerization systems can be attributed to the nature of the surfactant used during polymerization. As shown in Figure 4, the primary difference between the two surfactants lies in their hydrophobic tail. MO consists of a long-chain hydrocarbon whereas PDMS-PEO block copolymer consists of a polydimethylsiloxane backbone.

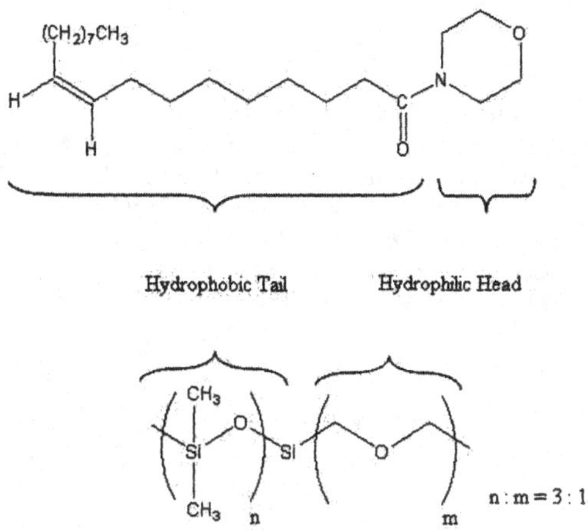

Figure 4. Molecular structure of top: Morpholine Oleate, bottom: PDMS-PEO block copolymer

The length of the PDMS backbone is much greater than the hydrocarbon chain in MO, as shorter surfactant tail would constitute to smaller micelles in the emulsion, this contributes to the observed size variation of MO-based PMMA particles. The flexible nature of the hydrocarbon chain in MO may also contribute to the apparent high polydispersity of the PMMA particles by forming labile micelles.

The PDMS backbone consists of two methyl groups per siloxane unit, whereas MO consists of repeated CH_2 units. The two methyl groups on the siloxane backbone contributes steric hindrance to rotational, torsional and bending modes along the hydrophobic tail of PDMS. This may explain the observed heat resistance of PDMS-based PMMA as the rigid hydrophobic tail remains relatively untangled within the micelle and allowed the PDMS-PEO surfactant to pack closely together. Effectively, PDMS-PEO surfactant formed a rigid spherical scaffold during micellar formation which ultimately provided anchorage to the polymerized particles, preventing the diffusion of polymeric chain segments during subsequent heat treatments.

CONCLUSION

Traditionally, the degree of cross linkage primarily governed the physical and thermal properties of any polymer. However, this study has shown that polymer latex properties may vary dramatically by a single change in the surfactant system used during polymerization. Additional studies involving the use of polymers with greater rigidity and steric hindrance as surfactants are necessary to understand the mechanisms involved in the observed heat-resistance.

REFERENCES

1. J.C. Padget, *J Coatings Technol.*, **66**, 89 (1994)
2. M. Sivakumar, R. Panduranga., *J. Appl. Polym. Sci.*, **83**, 3045 (2002)
3. T. Guo et al., *J Non-Crystalline Solids*, **353**, 2893 (2007)
4. W. D. Harkins, R. S. Stearns, *J. Chem. Phys.*, **14**, 215(1947)
5. J.L Keddie, *Mat. Sci. and. Eng.*, **21**, 101 (1997)
6. Millipore™, *Millipore - Product Detail*, viewed 01 Nov 2008, http://www.millipore.com/catalogue/item/sims000ge

Mater. Res. Soc. Symp. Proc. Vol. 1134 © 2009 Materials Research Society   1134-BB05-13

Effective Gas Sensing Using Quartz Crystal Microbalance Modified With Polymer Nanobrushes

Mutsumi KIMURA[1,2], Takashi MIHARA[3], and Tadashi FUKAWA[2]

[1]Collaborative Innovation Center for Nanotech FIBER, Shinshu University, Ueda 386-8567, Japan

[2]Department of Functional Polymer Science, Faculty of Textile Science and Technology, Shinshu University, Ueda 386-8567, Japan

[3] Olympus, Tokyo, Japan

ABSTRACT

We demonstrated weight-detectable quartz crystal microbalances (QCM) modified by various kinds of polymer brushes and investigated their sensing abilities for various organic gas molecules. These polymer brushes were attached onto the gold surface of QCMs through the combination of self-assembling monolayer formation and atom transfer radical polymerization (ATRP). ATRP polymerizations of three vinyl monomers from the initiator-modified QCMs resulted in the highly dense polymer brush layer with 40-60 nm thickness. The responses on polymer brush coated QCMs exhibited reversible frequency changes for various gas species and the frequency changes depend on the chemical structures of polymer brushes as well as gas species. We believe that coating with polymer brushes will provide a new approach for the highly effective and selective molecular sensing nanolayers for atmospheric gas molecules.

INTRODUCTION

Analyses of volatile organic compounds (VOCs) give us important information for safe and health. Continuous exposure to VOCs including formaldehydes, toluene and xylene cause serious problems for health. Specific trace vaporous compounds evaporate from dangerous bombs and drugs. Development of precise small sensors for VOCs leads to key technologies for direct monitoring on safe and health information around us. Swager et al. have developed highly sensitive molecular sensor for TNTs by using fluorescence quenching of conjugated polymers [1]. The appropriate nano-scaled space around fluorescent conjugated polymers allowed a rapid penetration of vaporous TNTs within solid membrane and the rapid penetration of vaporous electron-accepting analytes resulted in the efficient fluorescence quenching of conjugated polymers. Molecular imprinting has also investigated as selective molecular recognition devices through the control of intermolecular interactions and the size matching of analyte molecules in polymer matrixes [2]. These nanostrutures enhance sensitivity and selectivity of molecule-based sensors. In this study, we demonstrate VOCs sensors by the attachment of polymer nanobrush with the surface of quartz crystalline microbalance (QCM) (scheme 1).

Polymer coatings are attractive for a variety of applications ranging from protection layers in devices to responsive sites of sensors. In the latter case, polymers are versatile owing to their specific interaction with other molecules and their applicability for various different sensing principles [3]. When they are exposed to gases or vapors, polymers change their optical, electronic and mechanical properties. These changes provides us outputs to determine cencentration and species of gas molecules and these outputs were converted into the other signals such as weights, electroconductivities and capacitances of polymer films. In this context, polymer coated quartz crystal microbalances (QCMs) and electrodes have been fabricated and investigated their abilities for gas sensing [4]. Commonly, the polymer was coated by the several techniques including vapor-deposition, spin-coating, electrophoresis, and electropolymerization. However, it is difficult to provide uniform nanostructures in molecular level because of their long flexible nature. Recently, the attractive approach to control nanostructures on the solid substrates has been developed by the combination of self-assembling monolayer formation (SA) and living radical polymerization [5]. At the first, the thiols were attached with the gold surface by the formation of covalently bond Au-S and the self-assembled monolayers (SAMs) were formed on the surface of substrates. The other side of thiols possesses the attachment units for polymer chains and the corresponding monomers are polymerized from the attachment unit on the surface of SAMs. The spacial arrangement of polymer chains is observed due to the close distance of grafted points. Hence, the polymers deform and stretch in the direction perpendicular to the surface like brushes [6]. Furtheremore, atom transfer radical polymerization (ATRP) have been developed by Matyjaszewski et al [7]. ATRP allows the reaction to be carried out in a controlled way to generate a living growth end during the polymerization, and can be used to obtain polymers with high molecular weight and low polydispersity index. ATRP from the suraface of substrates induces the condense polymer brush formation with controlled length of polymer brush. The combination of SA with ATRP can produces uniform nanostructures composed of densed polymer brushes on the solid surface. These methods have been applied to the polymer-brush patterns and nanostructured surface for the attachment of biomolecules [8]. In this study, we demonstrate the fabrications of wight-detectable sensors by the polymer-brush coating on QCMs through SA and ATRP. Polymer naostructures can be provides uniform high-responsible surface for adsorption of gas molecules and the adsorption on the polymer brushes may induce the rapid and precise detection of weights of atmospheric gas molecules. In this paper, polymer brushes were attached with the surfaces of QCMs, and the morphologies and chemical structures, and analyzed their sensing abilities for volatile organic compounds (VOCs) were characterized.

Scheme 1.

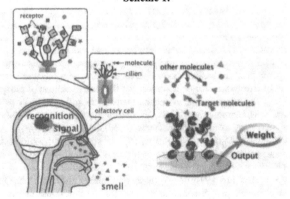

EXPERIMENT

Preparation of Polymer Brushes on substrates

The polymerizations of several vinyl monomers were carried out by the ATRP method from the initiator-modified substrates. Au-coated Si substrates were cleaned in piranha solution (3:1 (v:v) conc. H_2SO_4 and 30% H_2O_2) for 30 min, then were rinsed completely with DI (18.2 M Ω cm) water and dried under a stream of dry nitrogen [9]. SAMs of thiol having a initiator segment for ATRP were formed onto the cleaned gold surface by immersion into 1 mM toluene solutions of 13,13'-dithio-bis-(2-bromo-3-tridecanone) at 60℃ for 1 day. Polymer brushes prepared from the initiator-modified gold surface at 60 ℃ for 3 h in the presence of $FeBr_2$, triphenylphosphine and ethyl-2-bromoisobutyrate according to the literature method. After the polymerization, the substrates were washed with methanol and toluene to remove no-attached polymer chains.

Dimethylamino ethyl acrylate was used for the monomers for the polymer brushes on the substrates. Au-evaporated Si wafers and quartz crystal microbalances (QCMs) were used as substrates. The morphologies and chemical structures of polymer brushes were analyzed by SEM, AFM and IR-RAS using Au-evaporated Si substrates. Furthermore, the thickness of polymer brushes was estimated from the frequency increasing of QCMs after the polymerization and the polymer brush-modified QCMs were used for the adsorption properties of VOCs.

Evaluation of Sensing Abilities of QCM system

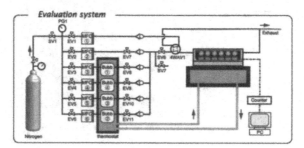

Figure 1. Evaluation system of VOC sensing

It is well established that a quartz crystal microbalance becomes a highly sensitive gas sensor when coated with the appropriate materials. The application of the QCM in gas sensors is based on interactions between the surface of a quartz crystal coated with a thin layer of sensitive material and the target molecules. These interactions result in detectable changes in the fundamental resonant frequency (Δf) of the resonator. When a small mass (Δm) is deposited or sorbed into a coating layer on the surface of the QCM, it induces a decrease in the resonant frequency, which is a function of (Δm). According to the Sauerbrey equation (1) [10], the deposition of a homogeneous coating promotes a shift of the natural resonant frequency of the crystal, which, for assumed negligible mass increase and rigid layer behavior, is described by the following linear relationship:

$$\Delta f = f_0 \, \Delta m \, / \, A \, \rho_q \, h_q \qquad (1)$$

where f_0 is the fundamental resonant frequency of the quartz crystal, ρ_q is the density of the quartz crystal, h_q is thickness of the quartz and A is the surface area of the electrode. The frequency changes of polymer-coated QCMs were monitored in the hand-made evaluation chamber having detecting counters (Agilent 53131A Universal Counter) at room temperature (Figure 1). The data were collected with an interval of 4 s and at 1 Hz and the frequency changes can be checked simultaneously. The polymer-coated QCMs were placed within the chamber and VOC containg gases were flowed. When VOC gases were injected, the collection of frequency was started at the same time. After reaching the equilibrium, the adsorption amounts of VOCs in films were estimated from the frequency change of QCM studies.

DISCUSSION

Polymer brushes were constructed by using ATRP polymerization of vinyl monomers from the initiator-modified surface according to the reported method (scheme 2).

Scheme 2. Synthetic approach of polymer brushes

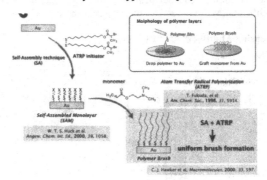

The FT-IR spectra of polymer brushes exhibited similar spectra of each parent polymers suggesting the successful polymerization by ATRP form the initiator-modified substrate surface. The morphologies of these polymer-modified surfaces were monitored by AFM and SEM. AFM images of polymer brushed Au substrate revealed a smooth surface with a thickness of approximately 40-60 nm. SEM image also indicates that the formations of polymer brush membrane on QCM surface were accomplished by this method.

Acetone, ethanol, and toluene were chosen as tested VOC sources and the concentration of these VOCs in the chamber were fixed at 4000 ppm. QCMs were coated with three polymer brushes were 50 nm thickness on average. When the acetone and ethanol were injected into QCM coated with polymer brushes, the decreasing of frequency value was observed. It suggests the weight increase of the polymer brush layer. The vapor sensing properties of polymer brushes on QCMs were investigated by measuring the frequency changes when the film was exposed to three VOCs vapors (Figure 2). We found that poly((2-dimethylamino)ethyl methacrylate) brush shows a strong affinity for every vapors than other polymer brushes and the response sequence of poly((2-dimethylamino)ethyl methacrylate) was ethanol > acetone > toluene, indicating the selectivity of polymer brush.

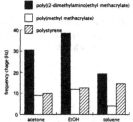

Figure 2. Sensor responses of QCM sensors modified with only three polymer brushes
to 4000 ppm acetone, ethanol, and toluene vapors at 20°C.

To demonstrate the moist capture ability of poly((2-dimethylamino)ethyl methacrylate) brushes, we also investigated the response to surrounding humidity (Figure 3). The equilibrium frequency differences of QCMs coated with poly((2-dimethylamino)ethyl methacrylate) brushes showed a good relationship to the humidity. We found that the QCMs coated with hydrophilic polymer brushes acts as an effective humidity sensor.

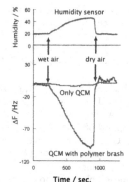

Figure 3. Frequency change of QCMs coated with poly((2-dimethylamino)ethyl methacrylate) brushes by exposition of humid air.

CONCLUSIONS

In summary, we have demonstrated the fabrication of weight-detectable sensors by the polymer-brush coating on QCMs through SA and ATRP. Sensing ability can control by the side chains of polymer brushes. Due to the different behavior of the polymer layers to tested gas species, it would be possible to detect the concentration VOCs and the recognition of gas species. We believe that supramolecular arrangement of polymer brushes onto the surface of QCMs provides a new approach to construct the highly effective and selective molecular sensing layers for the atmospheric gas molecules.

ACKNOWLEDGMENTS

This work was partially supported by a project for "Creation of Innovation Centers for Advanced Interdisciplinary Research Areas (Shinshu University)" in Special Coordination Funds for Promoting Science and Technology from MEXT of Japan.

REFERENCES

1. D. T. McQuade, A. E. Pullen, T. M. Swager, "Conjugated Polymer-Based Sensory Materials", *Chem. Rev.*, Vol. 100, pp. 2537-2574 (2000).
2. G. Vlatakis, L. I. Andersson, R. Müller, K. Mosbach, "Drug assay using antibody mimics made by molecular imprinting", *Nature*, Vol. 361, pp 645–647 (1993).
3. G. Harsanyi, "Polymer Films in Sensor Applications", Technomic Publishing Co., Inc., Lancaster (1995)
4. K. Nakamura, T. Nakamoto, T. Moriizumi, "Classification and evaluation of sensing films for QCM odor sensors by steady-state sensor response measurement", *Sens. Actuators B*, Vol. 69, No.3 pp.295–301 (2000)
5. R. Jordan, A. Ulman, J. F. Kang, M. H. Rafailovich, J. Sokolov, "Surface-Initiated Anionic Polymerization of Styrene by Means of Self-Assembled Monolayers", *J. Am. Chem. Soc.*, Vol.121, No.5, pp.1016-1022 (1999) b) Y. Tsujii, K. Ohno, S. Yamamoto, A. Goto, T. Fukuda, "Structure and Properties of High-Density Polymer Brushes Prepared by Surface-Initiated Living Radical Polymerization", *Adv. Polym.Sci.*, Vol.197,No.1 pp.1-45 (2006)
6. Milner, S. T. "Polymer Brushes", *Science*, Vol.251, pp.905-914 (1991)
7. K. Matyjaszewski, T. E. Patten, J. Xia, "Controlled / "Living" Radical Polymerization. Kinetics of the Homogeneous Atom Transfer Radical Polymerization of Styrene", *J. Am. Chem. Soc.*, Vol.119, No.4, pp.674-680 (1997)
8. Z. Bao, M. L. Bruening, G.L. Baker, "Rapid Growth of Polymer Brushes from Immobilized Initiators", *J. Am. Chem. Soc.*, Vol. 128, No.28, pp. 9056-9060 (2006) and R. R. Shah, D. Merreceyes, M. Husemann, I. Rees, N. L. Abbott, C. J. Hawker, J. L. Hedrick, "Using Atom Transfer Radical Polymerization To Amplify Monolayers of Initiators Patterned by Microcontact Printing into Polymer Brushes for Pattern Transfer", *Macromolecules*, Vol. 33, No.2 pp. 597-605 (2000)
9. R. L. Bunde, E. J. Jarvi, J. J. Rosentreter, "Piezoelectric quartz crystal biosensors", Vol. 46, pp.1223-1236 (1998)
10. G. Z. Sauerbrey, "Verwendung von Schwingquarzen zur Wägung dünner Schichten und zur Mikrowägung", *Z. Phys.*, Vol.155, pp. 206-208 (1959)

Mater. Res. Soc. Symp. Proc. Vol. 1134 © 2009 Materials Research Society   1134-BB05-21

Amplitude Interference Immune pH Sensing Devices Based on White Light Interferometry

C. R. Zamarreño, J. Goicoechea, I. R. Matias and F. J. Arregui
Sensor Research Laboratory, Electrical and Electronic Engineering Department, Universidad Pública de Navarra, Campus Arrosadía S/N, Edif. de los Tejos, Pamplona, NA 31006 SPAIN

ABSTRACT

We introduce here a fiber optic pH sensor based on white light interferometry which overcomes the major drawbacks of many optical fiber pH sensors such as their dependence to optical power variations or leaching of the indicators as well as the dependence of large coherence light sources. Here, communications optical fibers hold the basic sensing structure, which is fabricated using the Layer-by-Layer self electrostatic self assembly method and consists on the alternate deposition of the polymers poly(allylamine hydrochloride) and poly(acrylic acid) as the cationic and anionic monolayers respectively up to 50 bilayers. These nano-coatings operate as a Fabry-Perot Interferometric cavity of a few hundreds of nanometers suitable to be excited with a halogen white light source (FWHM ~300nm). In fact, the FPI-cavity acts as an intensity independent pH detector by monitoring the changes in the spectra waveform originated due to the changes in thickness of the coatings. Here, several measurements were performed with the fabricated sensors showing their robustness against changes of the optical power and sensitivity between 0.051 and 0.0187 pH units for the best and worst case respectively.

INTRODUCTION

Optical fiber sensors have experimented significant advances in the past decades as a result of the advantages of this kind of sensors (electromagnetic immunity, biocompatibility, multiplexation, small size, etc.), leading them to new and promising sensing devices [1]. In particular, optical fiber sensors are especially indicated in the monitoring of chemical and biological magnitudes, as far as such tasks can be easily measured by mean of optical phenomena like light absorption changes or fluorescence variations using the appropriate indicators [2]. Although these approaches are simple and cost-effective; they are mostly based in amplitude monitoring techniques, which can induce to measurement errors due to fluctuations in the intensity of the light source, leaching or bleaching of the indicators, etc. [3]. However, the amplitude measurements have been overcome with the development of several techniques that use wavelength-based measurements such as Plasmon resonance sensors [4] or interferometric sensors [5]. The main drawback of the wavelength-based sensors is that they typically use more complicated optical setups, or coherent light sources that increase significantly the cost of the systems.

In this work, the use of nanostructured coatings fabricated using the Electrostatic Self-Assembly Layer-by-Layer (LbL) technique allows the creation of high quality thin films with thicknesses of a few hundreds of nanometers [6] which can be used as white light interferometric cavities [7]. Here, it is fabricated and tested an optical pH sensor based on the interferometric response of pH sensitive nanostructured thin films which change their thickness depending on the pH of the surrounding medium [8]. This novel sensor, combines the advantages of the optical wavelength-based sensors with a cost effective white light interferometric setup.

EXPERIMENTAL DETAILS

All chemicals were purchased from Sigma-Aldrich Inc. and used without any further purification. The materials used in this work are the polymers *Poly(allylamine hydrochloride)* (PAH) and *Poly(acrylic acid)* (PAA) whose molecular structures can be seen in Figure 1. Standard communications multimode optical fibers (62.5/125μm core and cladding diameters respectively, purchased from Telnet-RI S.A.) were cleaved using a Fujikura Inc. cleaver and used as substrates. 10mM PAH and PAA aqueous solutions were adjusted at pH 5.0 by adding a few drops of HCl or NaOH when necessary. Thin films up to 50 bilayers were deposited using the LbL method and the PAH and PAA cationic and anionic solutions respectively. All aqueous solutions were made using deionized water with resistivity higher than 18.0MΩ·cm and a total organic content of less than 10ppb.

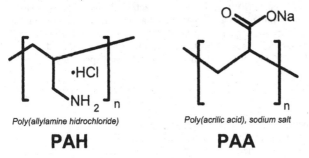

Figure 1. Chemical structures of the materials used in this work where PAH and PAA are the cationic and anionic polyelectrolytes respectively.

The optical response of the sensors was monitored using the setup shown in Figure 2, which consists in a DH-2000-H low-coherence (FWHM ~300nm) halogen white light source (Avantes Inc.), a standard communications multimode fiber coupler 50:50 (Telnet-RI S.A.), a PC with the appropriate software attached via USB to an OceanOptics USB4000 spectrometer and the sensing fiber tip.

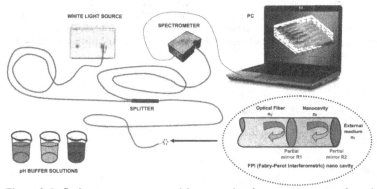

Figure 2. Reflecion-type setup arranged for measuring the spectral response from the sensor tips.

DISCUSSION

The optical path inside the polymeric nanostructure is lower than the coherence length of the halogen light source used. Thus, the interference phenomenon provides an oscillating reflectance response as the thickness at the end is varied [5]. The experimental data collected during the fabrication process of the 50 bilayers coating are shown in Figure 3 and confirm the expected oscillating response. The response in Figure 3 varies with the wavelength and also with the number of bilayers (thickness), which correspond with the interference pattern of a Fabry-Perot Interferometer (FPI).

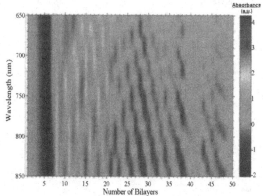

Figure 3. Experimental spectral response of every bilayer added to the structure up to 50 bilayers when it is fabricated at pH 5.0.

The interferometric properties of the polymeric cavity which change its thickness depending on the pH allows us to estimate the pH as a function of the spectral response of the device and therefore of the wavelength displacement of the interference fringes. The effect mentioned above is shown in Figure 4, where the absorption spectra are modified depending on the pH of the medium.

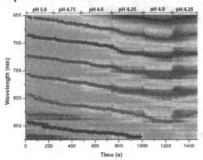

Figure 4. Monitored spectra of the sensors when they are immersed from pH 5 to pH 4 with a variation of 0.25 pH units and detected maxima (lines)

Furthermore, a peak detection algorithm can be used to position the maximum absorbance peaks for every single spectra, which correspond to the colored lines in Figure 4. The detection of the maximum absorbance peaks is used to monitor the displacements in wavelength originated by the changes in pH and characterize the dynamic response of the sensor. In Figure 5 is shown the evolution of the peaks detected between 725 and 775 nm with a maximum variation of 35 nm per pH unit. Moreover, the sensitivity varies from 0.051 to 0.0187 pH units per wavelength for the best and worst case respectively probing the feasibility of using this amplitude independent technique to measure pH.

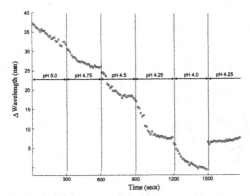

Figure 5. Variations in wavelength of the maximum absorbance peaks detected between 725 and 775 nm for each pH value.

124

CONCLUSIONS

To sum up, this manuscript proves the feasibility of producing wavelength-based pH sensors using a halogen light as the excitation source combined with the fabrication nanostructured thin films with the quality of varying their thickness depending on the pH of the medium where they are immersed. The results obtained in this work confirm the viability of using this technique to measure pH.

ACKNOWLEDGMENTS

This work was funded in part by the Spanish Ministry of Education and Science-FEDER TEC2006-12170/MIC Research Grant and Government of Navarre-FEDER Euroinnova Research Grants.

REFERENCES

1. B. Culshaw, J. Dakin, "Optical Fiber Sensors", *Artech House* (1997).
2. C. R. Zamarreño, J. Bravo, J. Goicoechea, I. R. Matias, F. J. Goicoechea, "Response time enhancement of pH sensing films by jeans of hydrophilic nanostructured coatings", *Sens. and Act. B* **129,** 138-144 (2007).
3. J. Goicoechea, C. R. Zamarreño, I. R. Matias, F. J. Arregui, "Minimizing the photobleaching of self-assembled multilayers for sensor applications", *Sens. and Act. B* **126,** 41-47 (2007).
4. J. Homola, S. S. Yee, G. Gauglitz, "Surface plasmon resonance sensors: review", *Sens. and Act. B* **54,** 3-15 (1999).
5. F. J. Arregui, I. R. Matias, Y. Liu, K. M. Lenahan, R. O. Claus, "Optical fiber nanometer-scale Fabry–Perot interferometer formed by the ionic self-assembly monolayer process", *Optics Letters* **24(9),** 596-598 (1999).
6. G. Decher, J.-D. Hong, "Buildup of ultrathin multilayer films by a self assembly process: II Consecutive adsorption of anionic and cationic bipolar amphiphiles polyelectrolytes on charged surfaces", *Ber Bunsenges. Phys. Chem.* **210/211,** 831-835 (1991).
7. J. Goicoechea, C. R. Zamarreño, I. R. Matias, F. J. Arregui, "Study on White Light Optical Fiber Interferometry for pH Sensor Applications", *Proceedings IEEE Sensors 2007,* 399-402 (2007).
8. K. Itano, J. Choi, M. F. Rubner, "Mechanism of the pH-Induced Discontinuous Swelling/Deswelling Transitions of Poly(allylamine hydrochloride)-Containing Polyelectrolyte Multilayer Films", *Macromolecules* **38,** 3450-3460 (2005).

Mater. Res. Soc. Symp. Proc. Vol. 1134 © 2009 Materials Research Society   1134-BB05-31

A Route Toward Wet Spinning of Single Walled Carbon Nanotube Fibers: Sodium Alginate - SWCNT Fibers

Vijoya Sa and Konstantin G. Kornev
School of Materials Science and Engineering, Clemson University, 161 Sirrine Hall, Clemson, South Carolina, 29634, USA, kkornev@clemson.edu

ABSTRACT

Polymeric fibers enriched with single-walled carbon nanotubes (SWCNT) are known as promising candidates for many applications in textile composites. In these applications, the fiber flexibility, toughness, as well as the fiber conductivity are important parameters. Ideally, one intends to make a fiber which would consist of an interwoven structure of nanotubes and polymer chains. This would make the fibers unique and distinguishable from any composite materials where the reinforcing elements are incorporated into a polymer matrix forming islands-in-a-sea-type structures. In popular method of fiber spinning based on injection of SWCNT dispersion into polymeric coagulation bath, the concentration of SWCNTs in the fibers is very difficult to control, and, typically, the polymer/SWCNT composition is unknown. As an alternative to this approach, we propose a new method based on conventional wet spinning technique. The idea is to exploit electrostatic assembling of SWCNTs coated with Sodium Dodecyl Sulfate (SDS). Many polymers important for applications, for example, alginate, chitosan, and polyacrylamide crosslink by ionic crosslinkers. Hence, if one manages to introduce SWCNTs-SDS into the network of these polymers, the resulting material is expected to be very strong and functional. We report on successful spinning of alginate fibers with carbon nanotube loading as high as 23 wt %. The Young's modulus increases up to 6.38 GPa at 23 wt % SWCNT loading. A transition from a composite structure (discrete SWCNTs embedded in alginate matrix) to a two-component polymer structure (SWCNTs incorporated into Alginate macromolecular network) is discussed using the results of mechanical and morphological characterizations.

INTRODUCTION

In the field of nanotechnology, carbon nanotubes (CNTs) take very important place due to their exceptional mechanical and electronic properties. Utilization of these properties is a challenge since the CNTs are commonly available in the powder form which assumes further processing to produce fibers, films or some other materials (1-7). Recent reports show that the addition of 0.1 – 3 wt % CNTs to polymers leads to significant improvement in the composite properties such as stiffness, strength, modulus, reactivity, conductivity, etc. Polyvinyl alcohol, polypropylene, polylactic acid, polyacrylonitrile, Nylon, and some other polymers were used as the matrix (8, 9).

Fiber spinning from nanotubes is the most challenging task in utilization of CNTs. Several methods were suggested to spin fibers from neat CNTs (10) and from polymer solutions (1, 11, 12). Neat CNT fibers and polymer-CNT fibers have different properties and hence different applications. In this paper we suggest new approach to spin fibers from polymers with high CNTs loading. The major challenge in wet spinning of fibers with CNTs is related to nanotube agglomeration which limits the existing methods to about 5 wt % of nanotube fraction in the dope (8, 9). To use these CNT-based fibers as artificial muscles, super capacitors, and support for biomedical devices, one needs to significantly increase CNTs loading. This would guarantee good electric conductivity along with enough flexibility of the fiber in question. Existing approaches to spin SWCNT-based fibers from solutions are mostly based on the method suggested in Ref. (1), when dispersion of carbon nanotubes is injected in polymer-containing coagulation bath (11, 12). Some modifications of this method have been reported in the literature, for example, spinning of CNT dispersion coaxially with the Polyvinyl alcohol (PVA) solution (13). In the method of Ref (1) or its modifications (11, 12) there is no practical way to have positive control on the amount of SWCNTs in the fiber: amount of collapsed polymer is unknown hence the CNTs concentration in the formed fiber varies from one experiment to the other. Here we report on successful formation of fibers from SWCNTs and Alginate through conventional wet spinning. This method allows one to significantly increase SWCNT loading, up to 23 % wt.

EXPERIMENTAL

Spinning process

High grade raw SWCNTs supplied by Nanoledge S.A., France, had been used in the experiments. Alginic acid sodium salt with high viscosity supplied by MP Biomedicals Inc. was used as a carrier. We used polymers with different molecular weight in the range between 120 KDa and 190 KDa. Calcium chloride used in the coagulation bath was extra pure crystals supplied by EMD Chemicals. Several trial experiments were done using different variations of alginate and SWCNTs compositions by varying the SWCNTs wt % in the fiber from 0.6 wt % to 23.1 wt %. The experimental procedure is as follows.

SWCNT dispersion was made in deionized water (DI water) using Sodium Dodecyl Sulfate as surfactant. SDS (1.7 wt %) was dissolved in DI water and then proper amount of SWCNTs (0.6 wt %) was added in the mixture and sonicated for 25 min to get homogeneous dispersion. In parallel, polymer solutions were prepared with different concentrations from 1.25 wt % to 2.0 wt % and mixed in proper ratios to achieve high SWCNT loading such as 0.6 wt %, 1.2 wt %, 2.4 wt %, 12 wt %, and 23.1 wt %. The SWCNT dispersion was stirred for 24 hr in the aqueous polymer solution. Then these colloidal solutions have been wet spun using 10 ml BD plastic syringe with the needle as a spinneret. The inner diameter of the needle was 0.85 mm. The extrusion rate was set at 123.2 ml/hr. The coagulation bath contained 15 w/v % aqueous solution of $CaCl_2$ and rotated at 33 RPM. After keeping them for 15 min in the coagulation bath, the fibers were transferred for overnight to another bath containing 3 w/v % aqueous solution of $CaCl_2$. These fibers were then completely dried in air and kept in refrigerator until testing. The spinning setup is shown in Fig. 1.

Fiber characterization

TENSILE PROPERTIES

Tensile testing of these fibers was done according to ASTM D3822 Single fiber break test method keeping 1 cm gauge length and 10 mm/min testing speed of Instron Tensile Testing Instrument. The tensile strength, elongation, and tensile moduli as well as the work of rupture were calculated from the obtained data.

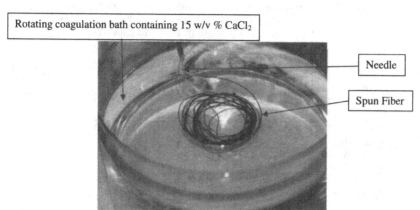

Rotating coagulation bath containing 15 w/v % CaCl₂

Needle

Spun Fiber

Fig. 1. Spinning process

SEM ANALYSIS

The surface morphology of the fibers as well as their internal structure have been analyzed with the Field Emission Scanning Electron Microscope (S4800; Hitachi, Tokyo, Japan). The samples were prepared by embedding the fibers in the epoxy matrix and then breaking them in liquid nitrogen to retain their inherent morphology and then coating with platinum for 40 seconds.

CONCEPT OF BONDING BETWEEN SODIUM ALGINATE-SWCNTS

In the mechanism of bonding of alginate to SWCNTs, sodium dodecyl sulfate, an ionic surfactant, plays very important role. It acts as a link between alginate and SWCNTs. In coagulation bath, calcium chloride dissociates to produce Ca^{+2}-ions. Making contact with sodium alginate, Ca^{+2}-ion replaces the Na^+-ion from the alginate and forms the calcium cage. This "egg-box model" was first proposed in Refs. (14-17). According to this model, along the alginate chain, the G-blocks adopt 2/1 helical conformation leading to the buckled regions. Within the cavities created by a pair of buckled G-sequences, Ca^{+2}-ion is coordinated forming an egg-box dimer (14). At the same time, the sodium caps in the SDS layers are also replaced by the Ca^{+2}-ions and, as a result, the Ca^{+2}-ions make calcium bridges between SWCNTs and alginate polymer. Ideally, if all ions, nanotubes, and polymeric chains work in unison, we should have a crystalline structure of SWCNTs - alginate in the fiber.

DISCUSSION

Using very simple industrially viable wet spinning method we formed sodium alginate fibers with the SWCNT loading as high as 23.1 wt %.

Table I. Tensile properties of the fibers with different SWCNT loadings

Sample	Diameter (μm)	Tensile strength (MPa) (Std. Dev.)	Tensile Modulus (GPa) (Std. Dev.)	Elongation (%)
Alginate fiber	49.3 – 56.2	210 (38)	2.53 (0.66)	17
0.6 wt % SWCNT – alginate fiber	48.9 – 52.2	220.2 (26)	4.93 (1.52)	19
1.2 wt % SWCNT – alginate fiber	43.0 – 49.1	259 (25)	5.51 (1.34)	13
2.4 wt % SWCNT – alginate fiber	56.6 – 57.8	152.9 (21)	3.93 (0.72)	18
12.0 wt % SWCNT –alginate fiber	67.1 – 71.2	202.9 (20)	4.37 (0.92)	15
23.1 wt % SWCNT –alginate fiber	61.1 – 66.7	260.3 (49)	6.38 (1.16)	14

As clearly seen from Table I and Fig 2, an increase of the SWCNT loading in the fiber from 0.0 wt % to 23.1 wt % leads first to an enhancement of the tensile modulus and the tensile strength and then to fiber weakening. Then the fiber becomes stronger again as the SWCNT concentration increases above ~ 2 wt%. The explanations of this unusual behavior lie in complex nanotube-nanotube and nanotube-alginate interactions. When the SWCNT loading is very low (less than 1.2 wt % SWCNT loading in our case), the nanotubes in the matrix are expected to be well dispersed, forming discrete systems of reinforcing elements. Thus, the behavior of this system is similar to the behavior of a fiber-reinforced composite: the composite becomes stronger with an increase of fiber concentration. When we further increase the nanotube loading, the nanotubes start to agglomerate during the fiber spinning: they form bundles which result in an effective decrease of the filler concentration in the composite, yet the effective diameter of the reinforcing element increases. The behavior then repeats the behavior of the fiber reinforced composites with larger fillers. Again, in accordance with composite mechanics, we have an increase of the tensile modulus. After a certain concentration, the bundles of nanotubes start to form its own network resulting in further increase of the fiber tensile properties.

In the fibers with 1.2 and 2.4 wt % SWCNT loading, neither the nanotubes nor the bundles of nanotubes are visible in the SEM micrographs. The bundles of nanotubes start to show up when the initial nanotube concentration reaches 12 wt %. In the fibers with 23.1 wt % SWCNT loading, the bundles are clearly seen in the cryo fractured sample, Figure 3. In Fig. 3 (a) the amount of visible nanotube bundles is small, yet visible. Hence, we can characterize this structure as a composite two-phase structure with discrete regions of SWCNT bundles in alginate polymer matrix. In Fig. 3 (b), corresponding to 23.1 wt % SWCNT loading, the SWCNT bundles are well spread over the alginate matrix. The structure looks like a single- phase multicomponent macromolecular structure. We have two interwoven components: SWCNT bundles and alginate chains. These SEM micrographs suggest that above some critical nanotube concentration, the SWCNT bundles form its own network making a better connection with the matrix. This macromolecular structure is able to share the load in a better proportion thus showing high modulus.

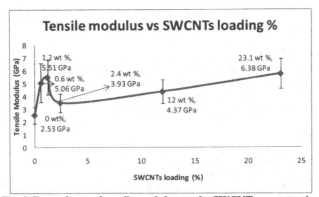

Fig. 2. Dependence of tensile modulus on the SWCNT concentration

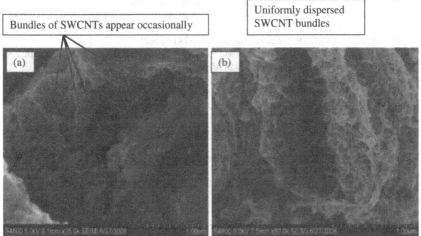

Bundles of SWCNTs appear occasionally

Uniformly dispersed SWCNT bundles

Fig. 3. SEM micrographs of cryo-fractured samples embedded in epoxy (a) 12 wt % SWCNT – alginate polymer fiber, (b) 23.1 wt % SWCNT –alginate polymer fibers.

CONCLUSIONS

Fibers with high loading of single walled carbon nanotubes were successfully produced and these fibers show high modulus and tensile strength. As high as 23.1 wt % loading of SWCNTs in alginate have been achieved. The fibers have been formed by simple industrially viable wet spinning process. Analyses of mechanical properties of these fibers and SEM characterization suggest that the nanotubes form a network interwoven into macromolecular network of alginate. Unusual concentration dependence of the Young's modulus with a well-

131

defined maximum supports this idea. SWCNT-alginate fibers are considered as good candidates for super-capacitors, artificial muscles, microelectrodes, supports for biomedical sensors, etc.

ACKNOWLEDGMENTS

We acknowledge the help of Dr. P. Brown, Dr. K. Stevens, and B. Ellerbrock at the initial stage of this work. We especially appreciate constant support of T. Andrukh, and D. Monaenkova. This research has been funded by the National Science Foundation, Award CMMI-0826067, and National Textile Center, Grant № M08-CL10.

REFERENCES

1. Vigolo, B., A. Penicaud, C. Coulon, C. Sauder, R. Pailler, C. Journet, P. Bernier, P. Poulin, *Science, 290*, 1331 (2000).
2. Baughman, R. H., Zakhidov, A. A., and de Heer, W. A., *Science, 297*, 787 (2002).
3. Kozlov, M. E., R. C. Capps, W. M. Sampson, V. H. ebron, J. P. Ferraris, R. H. Baughman, *Advanced Materials, 17*, 614 (2005).
4. Munoz, E., D. S., Suh, S. Collins, M. Selvidge, A. B. Dalton, B. G. Kim, J. M. Razal, G. Ussery, A. G. Rinzler, M. T. Martinez, R. H. Baughman, *Advanced Materials, 17*, 1064 (2005).
5. Poulin, P., B. Vigolo, P. Launois, *Carbon, 40*, 1741 (2002).
6. Dalton, A. B., S. Collins, J. Razal, E. Munoz, V. H. Ebron, B. G. Kim, J. N. Coleman, J. P. Ferraris, R. H. Baughman, *Journal of Materials Chemistry, 14*, 1 (2004).
7. Dalton, A. B., S. Collins, E. Munoz, J. M. Razal, V. H. Ebron, J. P. Ferraris, J. N. Coleman, B. G. Kim, R. H. Baughman, *Nature, 423*, 703 (2003).
8. Yang, J., T. Xu, A. Lu, Q. Zhang, Q. Fu, *Journal of Applied Polymer Science, 109*, 720 (2008).
9. Lee, S. H., M. W. Kim, S. H. Kim, J. R. Youn, *European Polymer Journal, 44*, 1620 (2008).
10. Zhang, X., Q. Li, Y. Tu, Y. Li, J. Y. Coulter, L. Zheng, Y. Zhao, Q. Jia, D. E. Peterson, and Y. Zhu, *Small, 3*, 244 (2007).
11. Neimark, A. V., S. Ruetsch, K. G. Kornev, P. I. Ravikovitch, *Nano Letters, 3*, 419 (2003).
12. Kornev, K. G., G. Callegari, J. Kuppler, S. Ruetsch, and A. V. Neimark, *Physical Review Letters, 97*, 188303(4) (2006).
13. Razal, J. M., J. N. Coleman, E. Munoz, B. Lund, Y. Gogotsi, H. Ye, S. Collins, A. B. Dalton, R. H. Baughman, *Advanced Functional Materials, 17*, 2918 (2007).
14. Y. P. Fang, Al-Assaf, S., Phillips, G. O., Nishinari, K., Funami, T., Williams, P. A., and Li, L. B., *Journal of Physical Chemistry B, 111*, 2456 (2007).
15. Grant, G. T., Morris, E. R., Rees, D. A., Smith, P. J. C., and Thom, D., Biological *Febs Letters, 32*, 195 (1973).
16. Morris, E. R., Rees, D. A., Thom, D., and Boyd, J., Chiroptical and Stoichiometric *Carbohydrate Research, 66*, 145 (1978).
17. Thom, D., Grant, G. T., Morris, E. R., and Rees, D. A., *Carbohydrate Research, 100*, 29 (1982).

Mater. Res. Soc. Symp. Proc. Vol. 1134 © 2009 Materials Research Society   1134-BB05-33

Hydrogen Generating Gel Systems Induced by Visible Light

Kosuke Okeyoshi and Ryo Yoshida
Department of Materials Engineering, Graduate School of Engineering, The University of Tokyo
7-3-1 Hongo, Bunkyo-ku, Tokyo 113-8656, Japan

ABSTRACT

Toward the hydrogen-energy society, researches on artificial photosynthesis have been done enthusiastically. But due to many problems, e.g., incomplete systems, low conversion efficiency, etc., they are still in the process of researches and development. In this study, we tried to achieve artificial photosynthesis system by using gel from viewpoints of systematic and molecular design. We prepared functional gels involving the electronic transmission circuit with functions of sensitizer, electron acceptor and catalysts. To verify each function in gel, we prepared three types of gels and measured the amount of generated H_2 gas for each gel system. In the gel system, hydrogen was generated more effectively than the solution system. Besides, by using thermo-responsive polymer as main chain, on-off control of H_2 generation was possible by changing temperature. All functional groups constructed in the gel network worked effectively and cooperated each other to generate H_2 gas.

INTRODUCTION

Biomimetic gels as information-converting materials are widely studied from the standpoint that life can be considered as an autonomous distributed cooperative system. In this study, hydrogen generating gel systems (sensitizer / acceptor / catalysts) [1] were designed by using cross-linked polymer network to generate hydrogen when visible light and water are supplied. Each functional group was introduced in thermo-responsive poly(N-isopropylacrylamide) (PNIPAAm) to operate electronic transmission circuit smoothly (Figure 1). By irradiating visible light to the gel, the functional groups constructed in the polymer network reduce water inside the gel and generate hydrogen. In the gel systems, the necessary components are not dispersed like solution systems in which colloidal particles become unstable due to high salt concentration and side reaction easily occurs. The Pt catalyst is trapped as a nanoparticle in the network, and the sensitizer and the acceptor units are arranged in the gel

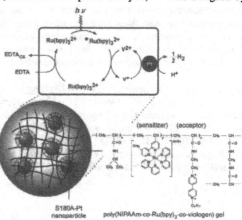

Figure 1. Mechanism of hydrogen generating systems using poly(NIPAAm-co-Ru(bpy)₃-co-viologen) gel containing Pt nanoparticles.

network organically. The Pt nanoparticles were prepared by using anionic surfactant as a protector. They can exist stably in the network without aggregation. To construct the gel system, the necessary components were introduced in the gel one by one, and each function was verified by visible light irradiation.

EXPERIMENT

Materials.

N-isopropylacrylamide (NIPAAm; Wako Pure Chemical Industries, Co., Ltd., Osaka, Japan) was purified by recrystallization from its toluene solution with hexane. Ruthenium(4-vinyl-4-methyl-2,2'-bipyridine)bis(2,2'-bipyridine)bis(hexafluorophosphate) (Ru(bpy)₃ monomer) was synthesized according to previous work.[2] N-β-acrylamidoethyl-N'-n-propyl-4,4'-bipyridinium bromide chrolide (viologen monomer) was synthesized according to the procedure reported before.[3] Other reagents were used without further purification.

Preparation of Gels.

Firstly, surfactant-modified Pt nanoparticle was prepared by alcohol reduction method [4] with chloroplatinic acid ($H_2PtCl_6 \cdot 6H_2O$)(Wako Pure Chemical Industries, Co., Ltd., JAPAN), reactive surfactant [5-6] (S180A)(Kao Co., JAPAN) as a protector and ethanol-water mixed solvent. After centrifugation, the colloidal solution was concentrated and dispersed in water. The diameter of surfactant-modified Pt nanoparticle was about 2 nm from the observation of transmission electron microscopy (TEM). Secondly, NIPAAm, Pt-colloidal suspension, Ru(bpy)₃ monomer, viologen monomer, and N,N'-methylenebisacrylamide as a cross-liker were dissolved in ethanol. Under nitrogen atmosphere for 20 min, 2,2'-azobis-2-2,4-dimethylvaleronitrile (ADVN) as an initiator were added to this pre-gel solution. The microgels were prepared by suspension polymerization using liquid paraffin as oil phase at 50 °C for 3 hours. After gelation, they were thoroughly washed to remove unreacted compounds. The diameter of the poly(NIPAAm-co-Ru(bpy)₃-co-viologen) microgel containing Pt nanoparticle at swollen state (20 °C) in water was about 300 μm from optical microscopy observation.

Measurements of H₂ generation for Gel Systems.

Three types of gel systems (G1 ~ G3); microgel suspension composed of microgels and reactants in the aqueous phase were prepared (Table I). These microgel suspensions were stirred sufficiently at 20 °C for 1 day, and then visible light was irradiated by using a halogen lamp (TECHNO LIGHT KTS-100RSV, Kenko). At given times, the absorption spectra was measured by UV-vis spectrophotometer (UV-2500PC, SHIMADZU), and the generated gas was collected and analyzed by gas chromatography (GC-8APT, SHIMADZU).

Table I. Components of the L and G systems.

System	Gel	Outer solution
L		EDTA/Ru(bpy)$_3^{2+}$/MV^{2+}/Pt
G1	PNIPAAm gel containing Pt NPs	EDTA/Ru(bpy)$_3^{2+}$/MV^{2+}
G2	Poly(NIPAAm-co-Ru(bpy)$_3$) gel containing Pt NPs	EDTA/MV^{2+}
G3	Poly(NIPAAm-co-Ru(bpy)$_3$-co-viologen) gel containing Pt NPs	EDTA

DISCUSSION

G1 system using PNIPAAm gel containing Pt nanoparticle.

When the G1 system was irradiated by visible light, absorption strength caused by MV$^+$· increased. We confirmed the abilities to sensitize and accept electron in the G1 system. Besides, by way of Pt nanoparticle in gel, hydrogen generated. This result indicates that water reduction by visible light in gel is possible. The amount of generated H$_2$ gas in the G1 system was compared with that of homogeneous solution system (L). Under the same concentration for all components, G1 generated H$_2$ gas more efficiently by 1.5 times than L. This is due to the concentration of Pt nanoparticle in gel and stabilization of Pt nanoparticles by polymer network.

G2 system using poly (NIPAAm-co-Ru(bpy)$_3$) gel containing Pt nanoparticle.

By irradiating visible light to the G2 system under stepwise temperature changes between 20 °C and 40 °C, the amount of generated H$_2$ gas changed remarkably. When the gel was at swollen state (20 °C) with hydrophilic environment, electronic transmission occurred smoothly and H$_2$ gas generated. On the other hand, when the gel was at shrunken state (40 °C) with hydrophobic environment, the decreased diffusivity in gel and skin layer of gel stopped electronic transmission and H$_2$ generation.

G3 system using poly (NIPAAm-co-Ru(bpy)$_3$-co-viologen) gel containing Pt nanoparticle.

By irradiating visible light to the G3 system, absorption strength caused by viologen unit increased. This result clearly shows that the electron was transmitted from Ru(bpy)$_3$ unit to viologen unit constructed in the network (Figure 2a, b). Moreover, H$_2$ gas generated, which means that electron transmitted effectively among three components constructed in the gel (Figure 2c). In future, by coexistence of catalyst for O$_2$ generation such as RuO$_2$, it would be possible to create gel systems performing complete artificial photosynthesis to generate both of hydrogen and oxygen when visible light and water are supplied. They are necessary for fuel cells and the gel systems are useful as a solar energy-converting system.

135

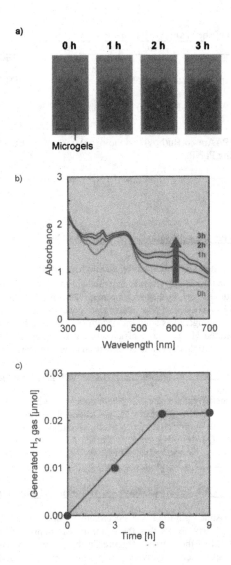

Figure 2. a) Color change of the G3 microgel suspension. b) Change in absorption spectra. c) H_2 generation for the G3 system at 20 °C under air atmosphere and light irradiation.

CONCLUSIONS

Firstly, Pt nanoparticle as a catalyst was immobilized in PNIPAAm gel. Hydrogen generated more effectively in the gel system than in the solution system. Secondly, Pt nanoparticle and Ru(bpy)₃ complex as sensitizer was immobilized and copolymerized in the gel. By changing temperature, the amount of generated H_2 gas could be controlled because the diffusivity of components in gel was changed remarkably. As well as Pt nanoparticle and Ru(bpy)₃ complex, viologen as acceptor was copolymerized in the gel. By irradiating visible light to poly(NIPAAm-co-Ru(bpy)₃-co-viologen) gel containing Pt nanoparticle, electron transmitted and hydrogen generated effectively.

ACKNOWLEDGMENTS

K. O. is grateful for the research fellowships of the Japan Society for the Promotion of Science for Young Scientists.

REFERENCES

1. K. Kalyanasundaram, M. Grätzel, *Angew. Chem. Int. Ed.* **1979**, *18*, 701.
2. R. Yoshida, T. Takahashi, T. Yamaguchi, H. Ichijo, *J. Am. Chem. Soc.* **1996**, *118*, 5134.
3. H. Kamogawa, H. Mizuno, Y. Todo, M. Nanasawa, *J. Polym. Sci., Polym. Chem. Ed.* **1979**, *17*, 3149.
4. N. Toshima, T. Takahashi, H. Hirai, *J. Macromol. Sci-Chem. A* **1988**, *25*, 669.
5. Y. Noguchi, K. Okeyoshi, R. Yoshida, *Macromol. Rapid Commun.* **2005**, *26*, 1913.
6. K. Okeyoshi, T. Abe, Y. Noguchi, H. Furukawa, R. Yoshida, *Macromol. Rapid Commun.* **2008**, *29*, 897.

Mater. Res. Soc. Symp. Proc. Vol. 1134 © 2009 Materials Research Society 1134-BB05-36

Preparation of Macroporous Poly(divinylbenzene) Gels Via Living Radical Polymerization

Joji Hasegawa[1], Kazuyoshi Kanamori[1], Kazuki Nakanishi[1], Teiichi Hanada[1], Shigeru Yamago[2]
[1]Department of Chemistry, Graduate School of Science, Kyoto University,
Kitashirakawa, Sakyo-ku, Kyoto, 606-8502, Japan
[2]Institute for Chemical Research, Kyoto University,
Uji, 611-0011, Japan

ABSTRACT

Macroporous cross-linked polymeric dried gels have been obtained by inducing phase separation in a homogeneous poly(divinylbenzene) (PDVB) network formed by organotellurium-mediated living radical polymerization (TERP). The living polymerization reaction of DVB with the coexistence of a non-reactive polymeric agent, poly(dimethylsiloxane) (PDMS), in solvent 1,3,5-trimethylbenzene (TMB) resulted in polymerization-induced phase separation (spinodal decomposition), and the transient structure of spinodal decomposition has been frozen by gelation. Well-defined macroporous monolithic dried gels with bicontinuous structure in the micrometer scale are obtained after removing PDMS and TMB by simple washing and drying. The properties of the macropores have been controlled by changing starting composition.

INTRODUCTION

The heterogeneous networks formation by the conventional free radical polymerization is widely applied to prepare porous polymeric materials (polymer monoliths) [1]. Using "porogen", which is usually poor solvent for the network-forming components, results in the substantial heterogeneity in the network, and thus phase separation of microgels in the porogen is induced. The resultant morphology is composed of aggregated-microgels (-particles) and the solvent (porogen). In this way the porous structure can be obtained after removal of the solvent. Polymer monoliths thus prepared are applied especially to liquid-phase separation or reaction media [2,3]. However, a fine tuning of pore properties such as pore size, pore volume and morphology is relatively difficult because the pores are formed in-between segregated microgel particles that aggregate at random. For the purpose of improved efficiency in such applications, the pore properties should be adequately controlled.

In this study, organotellurium-mediated radical polymerization (TERP) [4] is for the first time applied to prepare polymer monoliths based on spinodal decomposition. The TERP technique, as well as atom transfer radical polymerization (ATRP) and reversible addition-fragmentation chain transfer (RAFT) etc., is known as one of the versatile living radical polymerization techniques which affords well-defined linear polymers [5]. Using DVB as a monomer, poly(dimethylsiloxane) (PDMS) as a polymeric agent, and 1,3,5-trimethylbenzene (TMB) as solvent, phase separation behavior is investigated when changing the concentration of TMB or PDMS.

EXPERIMENT

Divinylbenzene (DVB) (80% mixture of isomers) was purchased from Sigma-Aldrich Co. (USA). The solvent 1,3,5-trimethylbenzene (TMB) was purchased from Kishida Chemical Co., Ltd. (Japan). Polydimethylsiloxane (PDMS) (trimethylsiloxy-terminated, molecular weight M_w = 9,000 - 10,000 (DMS-T22)) and 2,2'-azobis(isobutyronitrile) (AIBN) were purchased from Gelest, Inc. (USA) and Tokyo Chemical Industry Co., Ltd. (Japan), respectively. Ethyl-2-methyl-2-butyltellanyl propionate (BTEE) was kindly supplied from Otsuka Chemical Co., Ltd (Japan). All chemicals were used as received.

Given amounts of PDMS, TMB, and DVB were mixed in a glass tube in the listed order. After stirring for ca. 3 min, the resultant homogeneous solution was degassed by ultrasonication for 3 min, and then AIBN was dissolved followed by purging with nitrogen supplied using a stainless-steel needle through a silicone resin septum for 10 min. After BTEE was added through another needle, the starting solution was transferred to an ampule and kept at 80 °C to allow the decomposition of AIBN into free radicals. Simultaneously, the living radical polymerization was allowed to proceed at the same temperature. After reacted for 24 h, the resultant wet gels were washed with tetrahydrofuran (THF) to remove the solvent, phase separator, and unreacted monomers, followed by evaporative drying at 60 °C for 24 h. We confirmed by elemental analysis of the dried samples that all of the included PDMS was effectively removed by the washing with THF. For the mercury porosimetry, and nitrogen sorption measurements described below, the gels were heat-treated by raising the temperature from room temperature to 150 °C in 2 h and kept for 6 h in air.

The microstructures of the fractured surfaces of the dried gels were observed by SEM (JSM-6060S, JEOL, Japan). A mercury porosimeter (Pore Master 60-GT, Quantachrome Instruments, USA) was used to characterize the macropores of the heat-treated gels, while nitrogen adsorption–desorption (Belsorp mini II, Bel Japan, Inc, Japan) was employed to characterize the meso- and micropores of the gels. For mercury porosimetry, the pore size was characterized using the Washburn equation assuming a cylindrical shape for the pores. For nitrogen adsorption-desorption, the pore size distribution and surface area were respectively calculated by the BJH method using the adsorption branch of each isotherm and the BET method. The heat-treated samples were degassed at 110 °C under vacuum for at least 1 h prior to each measurement.

RESULTS & DISCUSSION

Macropores

By TERP, we obtained transparent or translucent PDVB wet gels without adding PDMS. The resultant wet gels turned into non-porous PDVB dried gels after evaporative drying. This implies a narrow distribution of molecular weight just before the gelation, and therefore the resultant PDVB networks are highly homogeneous in terms of the distribution of cross-linking points. The non-porous nature means no trace of phase separation in the course of polymerization of DVB without PDMS in the present case. With incorporating PDMS in the starting solution, porous PDVB gels with various morphologies of porous microstructure attributed to spinodal-type phase separation are obtained as reported previously [6].

Figure 1 shows the SEM images of representative PDVB gels prepared from various starting compositions as listed in Table 1. Here we define sample notations as x-y, where the numbers x and y are the mass of PDMS in mg and volume of TMB in mL in the starting solution, respectively. The amounts of BTEE and AIBN are fixed as 80 μL (1 mol% of DVB) and 0.029 g (0.5 mol% of DVB). In Figure 1, it is presented that the well-defined bicontinuous structure develops with increasing amount of PDMS in the same way as previously reported in nitroxide-mediated radical polymerization (NMP) systems [6]. With further increasing PDMS, the bicontinuous morphology transforms into that with aggregation of particles. This is obviously resulted from the breakage of continuous gelling skeletons driven by the interfacial energy [7]. With increasing TMB while fixing PDMS, the macroporous morphology becomes finer, suggesting TMB is good solvent that enhances mutual compatibility of PDVB and PDMS.

Table 1. Starting compositions and macropore morphologies

	DVB (mL)	TMB (mL)	PDMS (g)	AIBN (g)	BTEE (μL)	Macropore morphologies
600-8	5	8	0.600	0.029	80	no macropores
675-8	5	8	0.675	0.029	80	bicontinuous
725-8	5	8	0.725	0.029	80	bicontinuous
750-8	5	8	0.750	0.029	80	bicontinuous
775-8	5	8	0.775	0.029	80	bicontinuous
800-8	5	8	0.800	0.029	80	aggregate of particles
825-8	5	8	0.825	0.029	80	aggregate of particles
800-9	5	9	0.800	0.029	80	bicontinuous
800-10	5	10	0.800	0.029	80	bicontinuous
850-10	5	10	0.850	0.029	80	bicontinuous
925-10	5	10	0.925	0.029	80	aggregate of particles

Figure 1. SEM images of the dried PDVB gels.

The PDVB polymers become less soluble, i.e. a system becomes thermodynamically unstable, upon polymerization of DVB, and the originally single phase solution separates into PDVB-rich and PDMS-rich phases. This situation can be explained by the mean field theory [8], where the change in the free energy of mixing, ΔG becomes

$$\Delta G \propto RT \left(\frac{\phi_1}{P_1} \ln \phi_1 + \frac{\phi_2}{P_2} \ln \phi_2 + \chi_{12} \phi_1 \phi_2 \right). \tag{1}$$

Here, ϕ_i and P_i (i=1,2) denotes the volume fraction and degree of polymerization (DP) of component i, and χ_{12} shows the interaction parameter, and R and T are the gas constant and temperature, respectively. We neglect TMB here because it only acts as a diluting agent to reduce repulsive interaction between PDVB and PDMS. So we assume the component 1 and 2 as PDVB and PDMS, respectively, in a quasibinary system. With the progress of polymerization of DVB, i.e. when P_1 increases, the free energy of mixing ΔG also increases (note that ln ϕ_i is always negative). Phase separation takes place when ΔG becomes positive (polymerization-induced phase separation). It should be noted that, due to the large difference in solubility parameters (~9.1 $cal^{1/2} \cdot cm^{-3/2}$ for polystyrene-co-divinylbenzene and ~7.3 $cal^{1/2} \cdot cm^{-3/2}$ for PDMS) [9], there is almost no miscible compositional region in PS/PDMS blends except for oligomer-oligomer blends [10]. In other words, PDVB and PDMS are extremely immiscible due to the significant enthalpic disadvantage upon mixing; the contribution from the loss of the mixing entropy is rather small. According to de Gennes [11], in the course of polymerization, the correlation in-between DVB-derived species become stronger, which is equivalent to the increase of correlation in the course of physical cooling, i.e. decreasing the temperature. Thus, increasing chemical bonds can be equivalently treated as the physical cooling and termed as the chemical cooling.

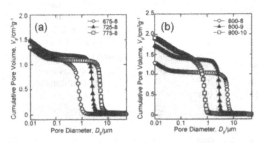

Figure 2. Cumulative pore volume and pore size distribution data obtained by mercury porosimetry. (a) Varying only PDMS concentration, and (b) varying only TMB concentration.

The macropore size and volume data obtained by mercury porosimetry are shown in Figure 2. It is confirmed from part (a) that only the pore size changes while the total pore volume holds almost constant when the amounts of PDMS is varied while keeping the solvent amount fixed. On the other hand, pore size becomes smaller and pore volume becomes larger with increasing amount of TMB as shown in part (b). Since TMB is considered to be the good solvent in the present system, it would be distributed in the both phases in the phase-separating and gelling solution (even though the distribution may not be even). But it eventually will be excluded from the PDVB-rich phase by syneresis due to the cross-linking and will be thrust into the PDMS-rich fluidic phase to determine the total pore volume. In this way, the total pore volume is mainly determined by the amount of TMB, and it should be noted that varying TMB also affects the pore size. It therefore is possible to obtain dried gels which have the similar pore size and different pore volumes by changing both amounts of TMB and PDMS adequately (simultaneously increase or decrease both PDMS and TMB).

Micro- and Mesopores

Inside the gel skeletons consisting macropores, smaller pores (hereafter referred as "skeletal pores") with various sizes are found by nitrogen sorption measurements. In Figure 3, it is given that the nitrogen adsorption-desorption isotherms of PDVB gels prepared with different concentrations of TMB in part (a) and PDMS in part (b).

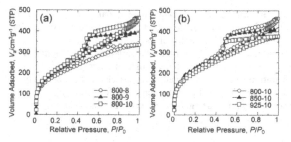

Figure 3. Nitrogen adsorption-desorption isotherms of the samples. (a) Varying only TMB concentration, and (b) varying only PDMS concentration.

In each isotherm, the continuous increase of adsorption from the low to high relative pressure range and the appearance of hysteresis loop between adsorption and desorption branches suggest the existence of pores with various sizes (meso- and macropores) inside the PDVB skeletons. The steep increase in adsorbed volume at the very low relative pressure region ($P/P_0 < 0.01$) can be attributed to the existence of micropores. Here, we denote the pores in accordance with the IUPAC recommendation; macropores are larger than 50 nm in diameter, mesopores are between 2 to 50 nm, and micropores are less than 2 nm [12]. In part (a), increasing amount of TMB makes the skeletal pore volume larger. Conversely, increasing amount of PDMS makes the pore volume smaller as depicted in part (b). Note that the increase in adsorption at higher relative pressures in the sample 800-10 is due to the small macropores formed by the early stage of phase separation. In the chemical cooling, being different from the physical cooling, the quench depth continues to become deeper by the continuous progress of polymerization [13,14]. Besides, in decreasing TMB or increasing PDMS, immiscibility becomes stronger and it would lead to the deeper quench. During this period, further phase separation (secondary phase separation or double phase separation) will take place inside the gelling phase (and also in the PDMS-rich fluidic phase) with the progress of further polymerization because of the high immiscibility between PDVB and PDMS as described above. For the gels with aggregate of particles such as 800-8 and 925-10, phase separation starts earlier enough prior to gelation. Additionally, since phase separation starts at when DP of DVB is relatively small and viscosity of the system therefore would remain low during phase separation. In such a system, secondary phase separation would well-proceed and the compositions in both phases become more close to the equilibrium ones. On the other hand, for the bicontinuous gels such as 800-9 and 850-10, the phase separation time is shorter, and moreover, viscosity during phase separation is higher. Secondary phase separation in both phases thus less-proceed and the both phase cannot follow the equilibrium compositions. That is, the structures formed by secondary phase separation are not fully relaxed and the larger amount of the skeletal pores remains as a result. Size of the skeletal pores may also be determined by secondary phase separation. They are however difficult to control because of the complex aspects of secondary phase separation: Quench depth and reaction rate as well as the phase separation time and viscosity largely affect the behavior of secondary phase separation [13,14]. It is worth noting that an additional reason for increasing skeletal pore volume with increasing TMB that can be confirmed from Figure 3 (a) is syneresis inside the skeletons considering from analogy with the first phase separation that forms macropores.

CONCLUSIONS

In the present paper, we have shown that phase separation (spinodal decomposition) can be induced in the homogeneous networks derived from organotellurium-mediated living radical polymerization (TERP) of DVB in the solvent TMB, and well-defined macroporous bicontinuous morphology can be consequently obtained after removing the solvent. With the addition of a polymeric agent PDMS that remains unreacted throughout the reaction, polymerization-induced phase separation takes place between PDMS-rich and PDVB-rich phases. The resultant porous dried gels possess two levels of pores; macropores formed by phase separation and skeletal pores by secondary phase separation inside the PDVB skeletons, which consists macropores.

ACKNOWLEDGMENTS

The present work was supported by the Grant-in-Aid for Young Scientists (B) (No. 20750177) from the MEXT, Japan, and was partly supported by a Grant for Practical Application of University R&D Results under the Matching Fund Method from New Energy and Industrial Technology Development Organization (NEDO), Japan. Also acknowledged is the Global COE Program "International Center for Integrated Research and Advanced Education in Materials Science" (No. B-09) of the Ministry of Education, Culture, Sports, Science and Technology (MEXT) of Japan, administrated by the Japan Society for the Promotion of Science (JSPS).

REFERENCES

1. E. C. Peters, M. Petro, F. Svec, J. M. J. Fréchet, *Anal. Chem.* **69**, 3646-3649 (1997).
2. F. Svec, *J. Sep. Sci.* **28**, 729-745 (2005).
3. M. R. Buchmeiser, *Polymer* **48**, 2187-2198 (2007).
4. S. Yamago, K. Iida, J. Yoshida, *J. Am. Chem. Soc.* **124**, 2874-2875 (2002).
5. W. A. Braunecker, K. Matyjaszewski, *Prog. Plym. Sci.* **32**, 93-146 (2007).
6. K. Kanamori, K. Nakanishi, T. Hanada, *Adv. Mater.* **18**, 2407-2411(2006).
7. T. Hashimoto, M. Itakura, H. Hasegawa, *J. Chem. Phys.* **85**, 6118-6128 (1986).
8. P. J. Flory, *Principles of Polymer Chemistry* (Cornell University Press: Ithaca, New York, 1971).
9. A. J. Ashworth, G. J. Price, *Macromolecules* **19**, 362-363 (1986).
10. T. Nose, *Polymer* **36**, 2243-2248 (1995).
11. P.-G. de Gennes, *Scaling Concepts in Polymer Physics* (Cornell University Press: Ithaca, New York, 1979).
12. K. S. W. Sing, D. H. Everett, R. A. W. Haul, L. Moscou, R. A. Pierotti, J. Rouquérol, T. Siemieniewska, *Pure & Appl. Chem.* **57**, 603-619 (1985).
13. T. Ohnaga, W. Chen, T. Inoue, *Polymer* **35**, 3774-3781 (1994).
14. T. Inoue, *Prog. Polym. Sci.* **20**, 119-153 (1995).

Mater. Res. Soc. Symp. Proc. Vol. 1134 © 2009 Materials Research Society   1134-BB05-38

Fabrication of Model Colloidal Systems With Tunable Optical Properties for Self-Assembly Studies

Adeline Perro[1], Guangnan Meng[2], Vinothan N. Manoharan[1,2]

[1] School of Engineering and Applied Sciences, Harvard University, Cambridge MA 02138, USA
[2] Department of Physics, Harvard University, Cambridge MA 02138, USA

ABSTRACT

We synthesize new types of micrometer-scale particles with engineered scattering properties. These particles are composed of small fluorescent polystyrene cores and large crosslinked poly(N-isopropylacrylamide) (polyNIPAM) shells. Immersed in water at ambient temperature, the shell of these particles is swollen and nearly index-matched to the surrounding fluid. In this state light scattering is dominated by the small polystyrene core, thereby eliminating multiple scattering from the suspension while still allowing the particles to be visualized under the optical microscope. Raising the temperature to ~35 °C leads to a decrease of the thermosensitive shell volume and an increase in scattering. We present some preliminary studies on the self-assembly of these colloids as characterized by confocal microscopy. These index-matched, aqueous particles might be used as tracers in microrheology experiments or as model colloids for real-space studies of phase behavior.

INTRODUCTION

The thickness of samples prepared for microrheology experiments or three-dimensional real-space studies of colloidal phase behavior is often limited by multiple scattering from the particles themselves. The particles need to be nearly index-matched in the suspension medium so that they do not scatter strongly, but they also need to be visible under optical or confocal microscopy. When the medium is water, it is difficult to satisfy both of these criteria. To overcome this limitation, we synthesize core-shell particles containing a large transparent shell associated with a small fluorescent core. The shell consists of a crosslinked polymer, poly(N-isopropylacrylamide) (polyNIPAM) that is highly swollen in water and nearly index-matched at room temperature. The scattering from these particles is therefore dominated by the small core, which we design to have a small scattering cross section. Thus even at close packing the particle centers are separated by more than the wavelength, and the scattering is weak while the particle centers remain visible under optical or confocal microscopy. We demonstrate that these colloids form nearly transparent crystals in water.

EXPERIMENTAL DETAILS

Fluorescent polystyrene seeds were synthesized using a recipe described elsewhere.[1] 50 mL of water were introduced in a 250 mL three-neck round-bottom flask equipped with a chilled condenser. The mixture was stirred at 300 rpm under nitrogen for 2 hours. 0.075 g of Sodium Dodecyl Sulfate (SDS, 99%, J.T. Baker), 0.4 g hexadecane (Fluka), 5 g of styrene (99%, Alfa Aesar) and 0.1 mg fluorescent dye (Nile Red, Aldrich) were then introduced. The emulsion was placed under ultrasound for 10 minutes to create a miniemulsion. The mixture was then heated to 70°C before the introduction of 0.05 g of potassium persulfate (KPS, 99%, Acros). Polymerization in miniemulsion droplets was carried out for 3h under magnetic stirring.

PolyNIPAM shell particles were synthesized according to procedures inspired from the literature.[2] A known amount of fluorescent polystyrene seed particles, 1.29 g of N-Isopropylacrylamide (NIPAM, 99%, Acros), 6 mg of cross-linker N,N'-Methylenebisacrylamide (MBA, Promega) and 50 mg acrylic acid (AAc, Sigma) were dissolved in 50 mL of water. The suspension was purged with nitrogen and the temperature increased to 70°C. 50 mg of potassium persulfate were used to initiate the polymerization. We used dynamic and static light scattering, turbidimetry and optical microscopy to characterize the final suspensions.

Crystals were formed using a depletion interaction.[3] 1 mL of small PolyNIPAM particles (60 nm in diameter, $C = 14.5$ g.L^{-1}) were added to 1 mL of PS-Poly(NIPAM-co-AAc) core-shell suspension ($C = 20.5$ g.L^{-1}) to induce the attraction. This mixture was then placed in a Vitro Com cell and sealed to prevent the suspension from evaporating. The crystals were imaged using a confocal microscope (Leica TCS SP5 Confocal Microscope).

DISCUSSION

Our PS-Poly(NIPAM-co-AAc) core-shell particles exhibit interesting optical properties due to their swollen shells. At ambient temperature, the shells are index matched with water, leading to transparent suspensions. Figure 1 shows an optical DIC (differential interference contrast) micrograph of such core-shell particles. The polystyrene beads are organized in an hexagonal close packing with a one micron interparticle distance. This regular arrangement is directed by shells that are invisible under the microscope because they are index matched with the water.

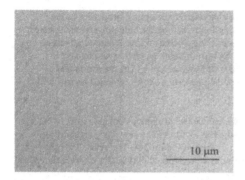

Figure 1: Optical microscopic observation of PS-Poly(NIPAM-co-AAc) core shell particles in DI water at room temperature.

We studied the size of the PS-Poly(NIPAM-co-AAc) shells as a function of temperature through dynamic light scattering (Figure 2). At ambient temperature (25°C), the shells are swollen by water, and the particle diameters are on the scale of a micrometer or so. Turbidity measurements show that the extinction coefficient of the core-shell particles is similar to that of the polystyrene core alone, indicating that the shell contributes negligibly to the scattering. An increase of the temperature above the lower critical solution temperature of the polyNIPAM (≈34°C) induced a dramatic change in the scattering. As the size of the shell decreases, the extinction coefficient increases. This phenomenon occurs because the polymer in the shell becomes more densely packed, as shown schematically in Figure 2.

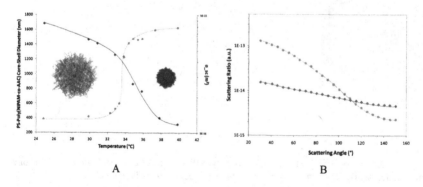

A   B

Figure 2: A) Evolution of the PS-Poly(NIPAM-co-AAc) core shell diameter (-) and extinction coefficient (-) as a function of the temperature, B) Static angular light scattering of PS-poly(NIPAM-co-AAc) core-shell particles at (-) 25°C and (-) 45°C, showing increase in forward scattering at elevated temperatures.

147

We used angular static light scattering to further characterize the optical properties of the particles. At ambient temperature, the scattering is dominated by the core polystyrene spheres, which exhibit nearly pure Rayleigh (isotropic) scattering. Increasing the temperature causes an increase in scattering cross section, as observed by turbidimetry, as well as an increase in forward scattering. Both of these are characteristic of Mie scattering. The transition was reversible leading to original materials where optical properties can be switched on and off [4].

To study the crystallization of these particles, we induced an attractive depletion interaction by adding small spheres to the suspension (see Methods above). After a few days, crystallites appeared on the glass surface of the sample cell, as seen in Figure 3 A. The crystallites showed strong Bragg diffraction, indicating that the particles were ordered over scales of several hundred micrometers.

To study in more details the organization of these colloids we used confocal microscopy, taking advantage of the fluorescent cores to define the position of each colloid center. Using particle tracking software,[5] we performed a full 3D reconstruction of the crystal. As seen in Figure 3 B, many regular planes of close-packed layers were present and the crystals as a whole showed face-centered cubic structures with stacking faults. The structure is similar to that observed for hard spheres.

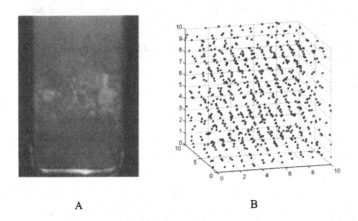

A   B

Figure 3: A) Picture of Bragg diffraction of PS-poly(NIPAM-co-AAc) crystal. Sample cell is one centimeter across. B) 3D reconstruction of cores alone (confocal analysis).

CONCLUSION

We have demonstrated the synthesis of micrometer-scale colloidal particles with controllable and tunable optical properties. At room temperature the particles are nearly index-matched in pure water, with fluorescent, weakly scattering cores that allow individual particles to be tracked using confocal or optical microscopy. These particles should prove useful as tracers in aqueous microrheology or as model colloids for studies of colloidal phase behavior.

ACKNOWLEDGMENTS

This research was funded by Egide (Lavoisier Fellowship, French Ministry of Foreign Affairs) and by the National Science Foundation under CAREER award number CBET- 0747625 as well as through the Harvard MRSEC under award number DMR-0213805. This work was performed in part at the Center for Nanoscale Systems (CNS), a member of the National Nanotechnology Infrastructure Network (NNIN), which is supported by the National Science Foundation under NSF award no. ECS-0335765. CNS is part of the Faculty of Arts and Sciences at Harvard University.

We would like to thank Jerome Fung for helpful discussions and Tom Kodger for assistance with the static light scattering measurements.

REFERENCES

1. S. Lelu, C. Novat, C. Graillat, A. Guyot and E. Bourgeat-Lami, *Polym Int*, **52**, 542-547 (2003).
2. H. Kawaguchi, K. Fujimoto and Y. Mizuhara, *Colloid Polym Sci*, **270**, 53-57 (1992).
3. A. D. Dinsmore, A. G. Yodh and D. J. Pine, *Physical Review E*, **52**, 4045-4057 (1995).
4. G. Wang, K. Kuroda, T. Enoki, A. Grosberg, S. Masamune, T. Oya, Y. Takeoka, and T. Tanaka, *Proc. Nat. Ac. Sci.*, **97**, 9861-9864 (2000).
5. http://physics.georgetown.edu/matlab/index.html - Matlab adaptation of IDL Particle Tracking software developed by David Grier, John Crocker and Eric Weeks.

Mater. Res. Soc. Symp. Proc. Vol. 1134 © 2009 Materials Research Society 1134-BB05-40

Temperature Responsive Wetting Between Hydrophilicity and Superhydrophobicity on Micropillar Arrays Grafted With Mixed Polymer Brushes

Yue Cui and Shu Yang
Department of Materials Science and Engineering, University of Pennsylvania, Philadelphia, PA 19104

ABSTRACT

Wetting behavior of surface and interface is of great importance for both fundamental science and practical applications. One of the major challenges is how to control the microscopic and nanoscopic roughness (surface chemistry, morphology and geometry) to contribute to the wettability and transition. Here, we report the temperature responsive wetting between hydrophilicity and superhydrophobicity on micropillar arrays grafted with mixed polymer brushes. Through sequential surface-initiated atom transfer radical polymerizations (SI-ATRP), a hydrophobic polymer, polystyrene (PS), and a thermal-responsive polymer, poly(N-isopropylacrylamide) (PNIPAAm) were grafted on epoxy micropillar arrays as mixed polymer brushes. At the room temperature, the surface of micropillar arrays exhibited hydrophilicity. When the temperature was increased to be above the lower critical solution temperature (LCST) of PNIPAAm, the PNIPAAm brushes collapsed and PS brushes dominated the pillar surface, exhibiting superhydrophibicity. The method we illustrate here offers a new approach to create a smart surface with dynamically tunable wetting behaviors by controlling both chemical heterogeneity and physical roughness. Such surface may find applications in microfluidics, micro- and nanofabrication of complex structures, crystal formation, and cell attachment, biosensors, actuator systems, and bioseparation.

INTRODUCTION

The wetting behavior of solid surfaces have attracted increasing attentions due to their potential important roles in the fields, including microfluidics [1], micro- and nanofabrication of complex structures [2], crystal formation [3], and cell attachment [4]. One of the major challenges is how to control the microscopic and nanoscopic roughness (surface chemistry, morphology and geometry) to contribute to the wettability and transition. Superhydrophilic and -phobic surfaces are fabricated by a combination of the effects of the surface roughness and chemical variation of the surface. A "smart" surface can be obtained from chemically homogeneous substrate, or surface that is modified by chemically different, responsive polymer chains from block copolymers, random copolymers, and polymer mixtures. Such a surface will respond to the subtle change of the environment, such as pH, temperature, light, and solvent quality [5, 6].

Mixed or binary polymer brushes have been created by grafting two distinct polymers one after the other on the surface by using either grafting-to or grafting-from approaches [7, 8].

Due to the limitation of the formation of polymer brushes with low grafting densities and low film thicknesses of the grafting-to approach, the grafting-from method is more attractive in terms of versatility, reliability, and controllability. Atom transfer radical polymerizations (ATRP)is of special interest because of its versatility, robustness, controllability, the living nature of the polymerization, and as a facile route to surface-grafted polymers, which are attractive because they can be used to tailor surface properties such as wettability, biocompatibility, adhesion, adsorption, corrosion resistance, and friction [9].

Here, we report the temperature responsive wetting between hydrophilicity and superhydrophobicity on micropillar arrays grafted with mixed polymer brushes. Through sequential surface-initiated ATRP, a hydrophobic polymer, polystyrene (PS), and a thermal-responsive polymer, poly(N-isopropylacrylamide) (PNIPAAm) were grafted on epoxy micropillar arrays as mixed polymer brushes or diblock copolymer brushes. At the room temperature, the surface of micropillar arrays exhibited hydrophilicity. When the temperature was increased to be above the lower critical solution temperature (LCST) of PNIPAAm, the PNIPAAm brushes collapsed and PS brushes dominated the pillar surface, exhibiting superhydrophobicity.

EXPERIMENT

N-isopropylacrylamide (NIPAAm) was purchased from ACROS Chemical and was recrystallized in n-hexane before usage. a-Bromoisobutyryl bromide and aminopropyl trimethoxysilane (APTES) were from Fluka (Switzerland). N,N,N',N',N''-pentamethyldiethylenetriamine (PMDETA), CuBr, and styrene (99%) were purchased from Sigma-Aldrich (USA). The master of micropillar arrays was kindly provided.

Epoxy micropillar arrays were treated with plasma oxidation, followed by the shaking in 5% APTES in anhydrous ethanol for overnight before being immersed in bromoisobutyryl bromide (0.1 ml in 10 ml of anhydrous ethanol and 0.15 ml of TEA) at 0 °C for 60 min. For the SI-ATRP of styrene to form PS, styrene (10 ml) and CuBr (55 mg) was degassed by nitrogen for half an hour before injecting the PMDETA (0.24 ml). The polymerization was performed at 100 °C for 4 hours.

The remaining active sites on PS grafted micropillars were quenched by soaking in 0.2 M sodium acetate solution (in 60% DMSO and 40% water) for 30 min, followed by backfilling of bromoisobutyryl bromide (1 ml in 10 ml of anhydrous hexane and 0.15 ml of TEA at 0 °C for 60 min. The PS grafted epoxy micropillars were dissolved in 10 mL of MeOH/H_2O (5: 5 / v:v) containing NIPAAm (1 g) and CuBr (32 mg) and degassed for 30 min before the injection of PMDETA (0.14 mg).

Static contact angles were measured with a Tantec contact angle meter (model CAM Micro) at room temperature. Contact angles were averaged from at least three different spots for each substrate. Scanning electron microscopy (SEM) images were obtained on FEI Strata DB235 Focused Ion Beam under voltage of 5 keV. A spot size of 3 was chosen for high resolution and detailed topographical information.

RESULTS AND DISCUSSION

As shown in Figure 1, the micropillar arrays fabricated by soft lithography were treated with oxygen plasma (Fig. 1a), followed by immersion in APTES to form the amino functional group on the surface, which were further immersion in bromoisobutyrl bromide to generate the 2-2-bromoisobutyrate-functionalized surface (Fig. 1b). In the nitrogen protecting atmosphere, in the presence of monomer NIPAAm, and catalyst PMDETA and CuBr, PNIPAAm was synthesized through ATRP (Fig. 1c). Due to the presence of active group bromide on the top of chain length of PNIPAAm, sodium acetate was used to quench the active site on PS for ATRP to avoid the formation of copolymer (Fig. 1d). The PS grafted micropillar arrays were further immersed in bromoisobutyrl bromide to graft the initiator for the synthesis of the second polymer PNIPAAm (Fig. 1e). With different reaction time for ATRP of NIPAAm, the chain length of PNIPAAm was different to exhibit different surface properties. When the temperature was below the lower critical solution temperature (LCST) of PNIPAAm, the surface of micropillar arrays exhibited hydrophilicity (Fig. 1f). When the temperature was increased to be above the lower critical solution temperature (LCST) of PNIPAAm (e.g. 50 °C), the PNIPAAm brushes collapsed and PS brushes dominated the pillar surface, exhibiting superhydrophibicity (Fig. 1g).

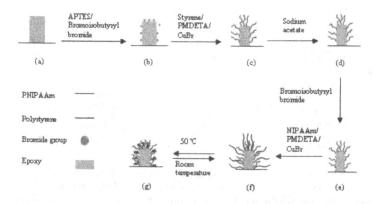

Figure 1. Schematic illustration for grafting mixed polymer brushes on micropillar arrays. (a) Oxygen plasma treated epoxy micropillar arrays (b) APTES and bromoisobutyrl bromide functional micropillar arrays (c) ATRP of styrene to form PS on micropillar arrays (d) Quench the active site on PS on micropillar arrays (e) Bromoisobutyrl bromide functionalization of on micropillar arrays (f) ATRP of NIPAAm to form the second polymer brushes PNIPAAm on micropillar arrays (e) Mixed polymer brushes grafted micropillar arrays at 50 °C.

The water contact angels for the micropillar arrays grafted with mixed polymer brushes were investigated. The SEM image of the micropillar arrays is shown in Fig. 2j. The oxygen plasma treated micropillar arrays exhibited superhydrophicility with a water contact angle of 0 °. After immersion in bromoisobutyrl bromide, the surface property became hydrophobicity of ~110 °. With the grafting of PS, a hydrophobic polymer, the surface became superhydrophobicity

(Fig. 2c). The remaining active sites on the polystyrene grafted micropillars were quenched by soaking in sodium acetate solution (Fig. 2d), followed by backfilling of bromoisobutyryl bromide (Fig. 2e), the surface kept superhydrophobicity with similar water contact angle as that just after grafting the PS brushes. The PS grafted epoxy flat substrate was further grafted with the second polymer brushes of PNIPAAm with ATRP. With different reaction time for ATRP, the surface exhibited different surface properties due to grafting with PNIPAAm with different chain lengths: e.g. with a reaction time of 5 min, the water contact angle increased from ~110 ° (Fig. 2f) to ~130 0 ° (Fig. 2g); with a reaction time of 30 min, the water contact angle increased from ~90 ° (Fig. 2h) to ~110 ° (Fig. 2i). This effect is explained by the competition between intermolecular and intramolecular hydrogen bonding below and above the LCST.

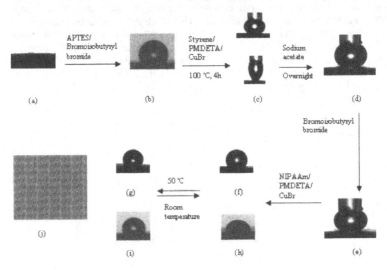

Figure 2. Water contact angles of micropillar arrays grafted with mixed polymer brushes. (a) Oxygen plasma treated epoxy micropillar arrays (b) APTES and bromoisobutyrl bromide functional micropillar arrays (c) ATRP of styrene to form PS (d) Quench the active site on PS (e) Bromoisobutyrl bromide functionalization (f) Mixed polymer brushes with PNIPAAm grafted for 5 min at room temperature (g) Mixed polymer brushes with PNIPAAm grafted for 5 min at 50 °C (g) Mixed polymer brushes with PNIPAAm grafted for 30 min at room temperature (i) Mixed polymer brushes with PNIPAAm grafted for 30 min at 50 °C (j) SEM image of the original micropillar arrays.

The water contact angles for flat substrates were also investigated, as shown in Figure 3. With the oxygen plasma treated, the flat epoxy surface exhibited hydrophilicity of 0 ° (Fig. 3a). With the treatment of APTES and bromoisobutyrl bromide for grafting the initiator, the water contact angle of the surface became ~90 ° (Fig. 3b). After grafting PS on the flat surface, the surface exhibited hydrophobicity of ~105 ° (Fig. 3c). The remaining active sites on the PS grafted flat surface were quenched by soaking in sodium acetate solution (Fig. 3d), followed by backfilling of bromoisobutyryl bromide (Fig. 3e), the flat surface exhibited hydrophobicity with similar water contact angle as that just after grafting the PS brushes. The PS grafted epoxy flat substrate was further grafted with the second polymer brush of PNIPAAm with ATRP. With the temperature increasing from room temperature (~22 °C) to 50 °C, the mixed polymer brushes grafted surface exhibited hydrophilicity with a water contact angle of ~65 ° to ~95 °. Due to the surface of being flat which lacks roughness, the wetting transition on the flat substrate was not as significant as that on micropillar arrays.

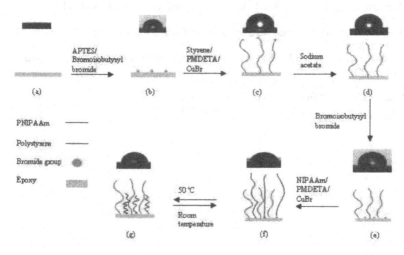

Figure 3. Schematic illustrations and water contact angles of flat surface grafted with mixed polymer brushes. (a) Oxygen plasma treated epoxy surface (b) APTES and bromoisobutyrl bromide functional surface (c) ATRP of styrene to form PS (d) Quench the active site on PS (e) Bromoisobutyrl bromide functionalization (f) Mixed polymer brushes with PNIPAAm grafted for 30 min at room temperature (g) Mixed polymer brushes with PNIPAAm grafted for 30 min at 50 °C.

CONCLUSIONS

We have demonstrated the temperature responsive wetting between hydrophilicity and superhydrophobicity on micropillar arrays grafted with mixed polymer brushes, which were created by SI-ATRP. The method we illustrate here offers a new approach to create a smart surface with dynamically tunable wetting behaviors by controlling both chemical heterogeneity and physical roughness. Such surface may find applications in microfluidics, micro- and nanofabrication of complex structures, crystal formation, and cell attachment, biosensors, actuator systems, and bioseparation.

REFERENCES

1. R. Seemann, M. Brinkmann, E. J. Kramer, F. F. Lange and R. Lipowsky, *Proc. Nat. Acad. Sci. USA* **102**, 1848 (2005).
2. G. S. Ferguson, M. K. Chaudhury, G. B. Sigal and G. M. Whiteside, *Science* **253**, 776 (1991).
3. J. Aizenberg, A. J. Black and G. M. Whitesides, *Nature* **398**, 495 (1999).
4. M. Mrksich, C. S. Chen, Y. N. Xia, L. E. Dike, D. E. Ingber and G. M. Whitesides, *Proc. Nat. Acad. Sci. USA* **93**, 10775 (1996).
5. M. Lemieux, D. Usov, S. Minko, M. Stamm, H. Shulha and V. V. Tsukruk, *Macromolecules* **36**, 7244 (2003).
6. M. Motornov, S. Minko, K. J. Eichhorn, M. Nitschke, F. Simon and M. Stamm, *Langmuir* **19**, 8077 (2003).
7. D.J. Li, X. Sheng, B. Zhao, *JACS* **127**, 6248 (2005)
8. B. Zhao and T. He, *Macromolecules* **36**, 8599 (2003).
9. K Matyjaszewski and J.H. Xia, *Chem. Rev.* **101**, 2921 (2001).

Mater. Res. Soc. Symp. Proc. Vol. 1134 © 2009 Materials Research Society   1134-BB05-41

Materials Properties of Polymer Blends of Poly(3,4-
ethlenedioxythiophene)/Poly(styrenesulfonate)/N-Methyl-2-pyrrolidinone
(PEDOT:PSS:NMP) and Polyvinyl Alcohol (PVA)

Chang-hsiu Chen[1], John LaRue[2] and Richard Nelson[3]
[1]Chemcial Engineering & Materials Science, University of California, Irvine, CA 92697, U.S.A.
[2]Mechanical & Aerospace Engineering, University of California, Irvine, CA 92697, U.S.A.
[3]Electrical Engineering & Computer Science, University of California, Irvine, CA 92697, U.S.A.

ABSTRACT

The structure material of microelectromechanical systems (MEMS) has been built with two polymer blends. It has tremendous applications in BioMEMS areas and optical electronic fields. Poly(3,4-ethlenedioxythiophene)/poly(styrenesulfonate) (PEDOT:PSS) is a high conductivity polymer. And a water soluble non-conductive synthetic polymer, polyvinyl Alcohol (PVA), leads to increase tensile strength, durability and flexibility. Mixtures were prepared and spin coated to form thin films. Detailed process of fabrication technique will be presented. The DC electrical properties of the thin films are studied using the four point probe method. Studies of electrical properties will lead materials to different applications. The film conductivities increased proportional to the increase in conducting polymer PEDOT:PSS content. Thermal treatment results will be also introduced in the article to demonstrate the stability of polymer blends. The conductivities did not change more than 2 times less after thermal or water/air treatment. In addition, N-Methyl-2-pyrrolidinone (NMP), a conductivity enhancer, was introduced and resulted in an increase of the conductivity of the polymer blends with the order of 100.

INTRODUCTION

Throughout the past two decades, the field of micro-electro-mechanical systems (MEMS) has grown rapidly[1]. It is recognized that future growth in the MEMS research depends in part on the development of additional materials including polymers[2]. Low cost, easy processing, excellent thermal stability, relatively high flexibility, ability to fabricate films in the range of 1 to 10 microns and low Young's Modulus provide the motivation for the development of polymer materials for MEMS. Therefore, conducting polymers have been widely used recently in MEMS, such as polymer actuators[3] and piezoresistivity cantilevers[4].

PEDOT (Poly 3,4-ethylenedioxythiophene)[5,6,7] is one of high conductivity conducting polymers, especially in OLEDs or transistors. PEDOT shows a high electrical conductivity (up to 500 S/cm) in the doped state. Also, it has good thermal and chemical stability[8]. However, PEDOT itself is insoluble and is difficult to process in solutions, so solubility was shown to be improved by using PSS (Poly(styrenesulfonate)) as the charge-balancing dopants during polymerization to yield PEDOT:PSS. (Figure 1.) The complex resulted in a water-dispersed polyelectrolyte system with good film forming properties, high conductivity (eg. 10 S/cm) and excellent stability. For example, films of PEDOT:PSS can be heated in air at 100°C for over 1000 hours with only a minimal change in conductivity[7]. Commercially, Bayer AG developed PEDOT:PSS with the trade name Baytron P where "P" stands for polymer.

However, due to lack of strong covalent bonds, PEDOT:PSS exhibits very poor durability and very brittle properties. To build 3D polymer structural MEMS devices, both electrical and mechanical properties have to be well developed. PVA, a water soluble synthetic polymer with excellent film forming, emulsifying, and adhesive properties[9] which has been a good selection toward our applications.

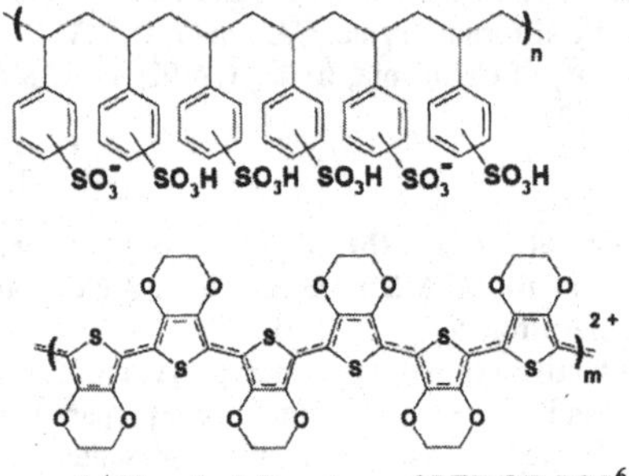

Figure 1. Chemical Structure of PEDOT:PSS[6]

PVA (Polyvinyl Alcohol, Figure 2.)[9] is a water soluble synthetic polymer. Normally, PVA is prepared by partial or complete hydrolysis of polyvinyl acetate to remove acetate groups. It has excellent film forming, emulsifying, and adhesive properties. It also has high tensile strength and flexibility. However, these properties are dependent on humidity. Specifically, at higher humidity levels more water is absorbed in the film and the higher levels of absorbed water, which acts as a plasticizer, leads to a reduction in the tensile strength but increases the elongation and tear strength. In summary, PVA when blended with PEDOT:PSS will lead to increased tensile strength, durability and flexibility of the film.

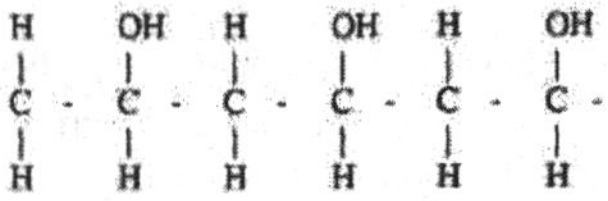

Figure 2. Chemical Structure of PVA[9]

Blending of polymers offers a means of combining useful and desired properties exhibited individually by the component polymers with a concomitant enhancement of important properties. In this study of PEDOT:PSS/PVA blend, PEDOT:PSS provides high conductivity[5], and PVA provides mechanical properties as a binder[9]. PEDOT:PSS blended with PVA, leads to thicker, more durable, and stronger films with better thickness uniformity. To achieve a controlled conductivity and strong mechanical structure, blends of PEDOT:PSS/PVA were prepared and their conductivity determined.

The present work reports the development of conductive polymer films for future applications in MEMS. The process to produce films from the PEDOT:PSS/PVA blend is presented and the conductivity and thickness of films for different percentages of the PEDOT:PSS and PVA are presented.

EXPERIMENT

PEDOT doped with PSS is commercially available in the form of a water dispersion (Baytron P V4071 from H.C. Starck). A high conductivity formulation NMP (N-Methyl-2-pyrrolidinone) and other various components specified by the manufacturer has been prepared[6]. PVA (98-99% hydrolyzed, high molecular weight material from Alfa Aesar) was used as received, and mixed in a 9 wt% water solution at higher temperature. PEDOT:PSS:NMP dispersion and PVA 9 wt% solution were blended in various percentage of PEDOT:PSS:NMP and PVA. Subsequently, polymer films were prepared by spin-coated PEDOT:PSS:NMP and PVA blends solutions on both oxidized silicon substrates and glass slides at 500 rpm for 30 seconds. These films were placed in a vacuum oven at 100°C for 6 hours to remove the solvent and prepared to perform further characterizations. All the films on the oxidized silicon wafer were then sectioned into 7mm by 40 mm samples for testing. The glass slides were not cut. Silver epoxy conducts, 1mm in width, were placed on the films using a stenciling process with 10mm separation between the electrodes.

The thickness of all of the films was measured using Digital Instruments Dektek 3 surface profiler. The electrical conductivity of the polymer films was measured with a Kelvin four-point probe technique using a LakeShore 7507 system with Hall Measurement System version 2.3.0 software. Thermal treatment was applied after 7 days and reheat to observe the change of conductivities.

DISCUSSION

Figure 3. shows the thickness measurement of spin coated films at 500 rpm spin rate, both on glass slides and the oxidized silicon wafers as a function of different weight percentages, wt%, of PEDOT:PSS:NMP and PVA. As shown in Figure 3, the film thickness decreases with increasing PEDOT:PSS:NMP weight percentage due to the fact that increasing the wt% of PVA increases the viscosity of the blend which leads to a thicker film for the fixed spin rate.

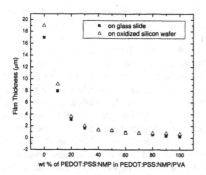

Figure 3. Film Thickness vs PEDOT:PSS:NMP/PVA weight percentage on both glass slides and oxidized silicon wafers

The conductivity as a function of PEDOT:PSS:NMP weight percentage is shown on Figure 4. In order to minimize potential contact effects, four-point probe testing was utilized. As expected, the conductivity increases with increasing PEDOT:PSS:NMP weight percentage. For example at a weight of 100% the film conductivity is about 10 S/cm on a glass substrate and 4.29 S/cm on an oxidized silicon substrate. In general, for the same weight percentage of PEDOT:PSS:NMP, the conductivity of the films on the glass slide varies for twice or more the conductivity of the film on the silicon substrate. Specifically at PEDOT:PSS:NMP weight percentages of 60%, 40%, and 20%, the conductivities on silicon wafers are respectively, 0.00968, 0.684, 2.99 S/cm and on glass slides are 0.0299, 1.89, 4.02 S/cm.

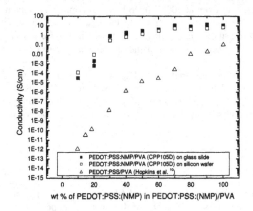

Figure 4. Conductivity as a function of wt% of PEDOT:PSS:NMP in a blend of PEDOT:PSS:NMP/PVA on both substrates and corresponding conductivity measurements of Hopkins and Reynolds[10] for PEDOT:PSS/PVA

Figure 4. also shows the reference data of Hopkins & Reynolds[10]. For all weight percentages of PEDOT:PSS, the conductivities measured in the present experiment are much greater than those of data. For example, the conductivity measured by Hopkins & Reynolds is 0.1 S/cm for 100wt% PEDOT:PSS. Contrast this with the conductivity of 10 S/cm measured at the same weight percentage in the current study. At lower weight percentages, the difference in the measure conductivity is considerably larger, like weight percentage of 40%, Hopkins and Reynolds reported a conductivity of 10^{-6} S/cm. Contrast this value with the values in the range of 0.3-0.8 S/cm were found in the current study. In order to verify the relationship of weight percentage and conductivities between current study and Hopkins & Reynolds, repeat of Hopkins & Reynolds results are also shown on Figure 5. The higher conductivities in the current study were obtained because of the conductivity enhancement agent NMP. NMP, an non-conductive organic solvent, is a conductivity enhancement with two or more polar groups on each molecule. The driving force for the conformational change of the PEDOT chains is the interaction between the dipole of one polar group of the organic compound and the positive charges on the PEDOT chains. Another polar group of the organic compound may form a hydrogen bond to PSS chains so that the organic compound can stay in the polymer film and in close proximity to the PEDOT chain[11]. In order to verify that NMP enhances the electrical conductivity, measurements with different amounts of NMP were performed. Figure 6. shows the conductivity increase with increasing amounts of NMP, indicating that NMP does enhance the conductivity. The PEDOT:PSS concentration was reduced with increasing NMP concentrations.

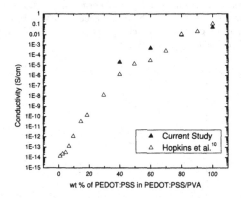

Figure 5. Conductivity as a function of wt% of PEDOT:PSS in PVA (without any NMP) for results obtained in the current study (solid triangles) and those of Hopkins and Reynolds[10] (white triangles)

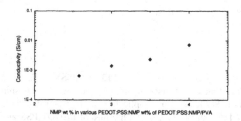

Figure 6. Conductivity with different amounts of NMP in wt%

Figures 1. and 2. show the chemical structure of PEDOT:PSS & PVA. PEDOT is a p-type electronic conductor. PSS is a negative ionic conductor. PEDOT doped PSS exhibits semiconductor properties. In our observation, the blends conductivity drops with PVA increases, however the mechanical strength looks stronger with PVA increases. When PVA is blended into PEDOT, the OH groups of PVA have been bonded by strong covalent bonds with the SO_3^- groups of PSS which reduce some negative charge of the PEDOT:PVA blend. So the more PVA concentration in the blends, the more SO_3^- groups of PSS will be bonded with OH groups of PVA, and eventually degrade the conductivity of blended polymers, due to the decreasing of conductive SO_3^- groups. On the other hand, since OH groups and SO_3^- groups have strong covalent bonds, more PVA concentration will result in stronger mechanical properties of films. More detailed mechanism of polymers conductivity could be investigated from impedance measurement of polymer films or chemical structures analysis.

Figure 7. shows the results of thermal treatment effects after 7 days in air atmosphere and reheat in 4 hours. Most of conductivities did not change a lot and the most decay of conductivity did not exceed half of original. Therefore it indicates the thermal and air atmosphere do not affect the conductivities and shows its physical durability and stability. Normally, the effects of conductivities could be contributed either moisture or air. These references[7,12] imply below 100°C, moisture would be the main reason of conductivities change and above 100°C, the conductivities could be attributed both moisture and oxygen influence.

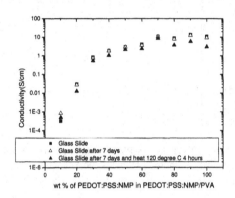

Figure 7. Thermal effect of PEDOT:PSS:NMP conductivities

CONCLUSIONS

Poly(3,4-ethylenedioxythiophene):poly(styrenesulfonate) (PEDOT:PSS) is a commercially available, conducting polymer that has a viscosity that is too low for the spin coating application of one micrometer and thicker films. Spin application of the water dispersed material leads to films that are very thin and non-uniform. Using a mixture of PEDOT:PSS and polyvinyl alcohol (PVA) solved this problem and resulted in films that can be applied by spin application and have useful conductivity. Use of the conductivity-enhancement material N-Methyl-2-pyrrolidinone (NMP) can increase the conductivity by several orders of magnitude. Films of the blended polymers with a PEDOT:PSS concentration equal to or greater that 50 wt% have a conductivity greater than 1 S/cm. And in addition, thermal treatment indicates the conductivities are constant and not effected by some heat or air atmosphere.

ACKNOWLEDGMENTS

The authors gratefully acknowledge Allen Kine for experimental assistance, Pai-chun Chang and Dr. Grace Lu for help in conductivity measurement and Dr. Jill Simpson from H.C.Starck for the discussion of properties of Baytron P.

REFERENCES

1. M. Gad-el-Hak, in MEMS Handbook-Introduction and Fundamental (CRC Press, Boca Raton, FL, 2002), p1-1~1-5
2. M. Gad-el-Hak, in MEMS Handbook-Design and Fabrication (CRC Press, Boca Raton, FL, 2002), p9-1~9-39
3. E. Smela, MRS Bulletin **33**, 197 (2008)
4. M. Lu, A. Bermak and Y. Lee, Micro Electro Mechanical Systems Conference 2007, 251 (2007)
5. S. Kirchmeyer and K.J. Reuter, Mater. Chem. **15**, 2077 (2005).
6. http://www.baytron.com/
7. L."Bert" Groenendaal, F. Jonas, D. Freitag, H. Pielartzik and J.R. Reynolds, Adv. Mater. **12**, 481 (2000)
8. J. Huang, P.F. Miller, J.C. de Mello, A.J. de Mello and D.D.C. Bradley, Synth. Met. **139**, 569 (2003)
9. C.A. Finch, in Polyvinyl alcohol; properties and applications (Wiley, London, 1973), p 17-65
10. A.R. Hopkins, J.R. Reynolds, Macromolecules **33**, 5221 (2000)
11. J. Quyang, Q. Xu, C. Chu, Y. Yang, G. Li and J. Shinar, Polymer **45**, 8443-8450 (2004)
12. I. Winter, C. Reese, J. Hormes, G. Heywang and F. Jonas, Chem. Phys. **104**, 207 (1995)

New Materials and Characterization III

Mater. Res. Soc. Symp. Proc. Vol. 1134 © 2009 Materials Research Society   1134-BB07-03

Stochastic System Identification of the Compliance of Conducting Polymers

Priam V. Pillai[1] and Ian W. Hunter[1]
[1] BioInstrumentation Lab., Department of Mechanical Engineering, Massachusetts Institute of Technology
Cambridge, MA 02139, U.S.A.

ABSTRACT

Conducting polymers such as polypyrrole, polythiophene and polyaniline are currently studied as novel biologically inspired actuators. The actuation mechanism of these materials depends upon the motion of ions in and out of the polymer film during electrochemical cycling. The diffusion of ions into the bulk of the film causes the dynamic mechanical compliance (or modulus) of the material to change during the actuation process. The mechanism of this change in compliance is not fully understood as it can depend on many different factors such as oxidation state, solvation of the film and the level of counter ion swelling. In-situ measurement of the dynamic compliance of polypyrrole as a function of charge is difficult since the compliance depends upon the excitation frequency as well as the electrochemical stimulus. Pytel et al [1] studied the effect of the changing elastic modulus in-situ at a fixed frequency. In this study we describe a technique to measure the compliance response of polypyrrole as a function of frequency and electrochemistry. A voltage input and a simultaneous stress input was applied to polypyrrole actuated in neat 1-butyl-3-methylimidazolium hexaflourophosphate. The stress input was a stochastic force with a bandwidth of 30 Hz and it allows us to compute the mechanical compliance transfer function of polypyrrole as function of the electrochemistry. Our studies show that the low frequency compliance changes by 50% as charge was injected into the polymer. The compliance changes reversibly as ions diffuse in and out of the film which indicates that the compliance depends upon the level of counter ion swelling.

INTRODUCTION

Electroactive conducting polymers are currently studied as materials that can have a wide range of applications including novel biologically inspired actuators, sensors, valves and pumps. They have a number of attractive properties such as being lightweight, inexpensive and are easy to mold and shape into a wide range of forms. Various electroactive polymers that are activated by ion diffusion are studied for use in a wide variety of applications. The actuation mechanisms in these materials are based on the diffusion of ions in and out of the polymer film. The films are capable of generating large strains of between 5-10% at low voltages (1 to 2 V) which make them ideal elements to build artificial devices that can mimic biological elements [2]. During electrochemical stimulus as charge is injected into the polymer there is an expansion of the polymer film that accounts for a large portion of the overall strain. However, there are a number of underlying mechanisms that can also govern polymer actuator behavior. Changing material properties during the actuation cycle is one such mechanism that has not been studied well. For instance, as ions diffuse into the polymer the compliance of the polymer also changes [1, 3-6]. If the polymer is held in tension then this changing compliance also contributes to the overall strain. This can be best understood from Equation 1 where ε_T is the total strain on the polymer, α is the strain to charge ratio, q is the charge injected into the polymer, Y is the compliance which can depend on the charge and excitation frequency and σ is the preload stress.

$$\varepsilon_T = \alpha q + Y(w,q)\sigma \qquad (1)$$

The overall strain is composed on a volumetric strain that is proportional to the charge injected into the polymer and a component that comes from the changing compliance. Many authors [3-6] have shown that the polymer compliance changes during the actuation cycle, however there are a number of discrepancies about how the compliance changes [1]. There is a need to develop standard techniques that can be used to measure the compliance that can be applied during the actuation process. In this paper we describe novel a technique to make measurements of the compliance of conducting polymers as a function of an electrochemical stimulus and excitation frequency. We do this by computing the compliance impulse response as a representation of the compliance transfer function of the polymer. The impulse response can be convolved with any input stress waveform to predict the resulting strain. We choose to use efficient stochastic system identification techniques to estimate the compliance impulse response of the polymer.

EXPERIMENTAL METHOD

Pyrrole (Aldrich 99%) was vacuum distilled before use. Polypyrrole was electrodeposited on a glassy carbon substrate at -40°C at a constant current density of $0.5A/m^2$. The deposition solution used was 0.05M pyrrole in 0.05M tetraethyl ammonium hexaflourophosphate (TEAPF6) in propylene carbonate. The resulting polypyrrole films were cut in 2mm × 10mm strips and tested using a custom built dynamic mechanical analyzer (DMA) [7]. This apparatus allows us to clamp the polymer in tension while applying an electrochemical stimulus using a three electrode cell. The stress generating DMA stage was used to generate a stochastic stress input and the corresponding stochastic strain was measured. The stress input consisted of a stochastic signal having a shaped power spectrum and Gaussian probability density function with mean amplitude of 8 MPa, a standard deviation of 0.5 MPa and a bandwidth of 30 Hz. Linear stochastic system identification techniques [8, 11] were used to estimate the parameters of the compliance impulse response function and the compliance frequency response function of polypyrrole.

$$R_{XX}h_{XY} = R_{XY} \quad \begin{bmatrix} R_{XX}(0) & R_{XX}(1) & \cdots & R_{XX}(n) \\ R_{XX}(1) & R_{XX}(0) & \ddots & \vdots \\ \vdots & \vdots & \ddots & \vdots \\ R_{XX}(n) & \cdots & \cdots & R_{XX}(0) \end{bmatrix} \begin{bmatrix} h_{XY}(0) \\ h_{XY}(1) \\ \vdots \\ h_{XY}(n) \end{bmatrix} = \begin{bmatrix} R_{XY}(0) \\ R_{XY}(1) \\ \vdots \\ R_{XY}(n) \end{bmatrix} \qquad (2)$$

To find the impulse response 40 point auto and cross correlation function estimates were made of the stress and strain and were used to fill the elements of Equation 1, where Rxx is a Toeplitz matrix of the autocorrelation function values, Rxy is the cross correlation function estimates and hxy is the of samples of the impulse response function to be determined. We solve for hxy using a Toeplitz matrix inversion implemented in MATLAB. The dynamic compliance frequency response function may then be determined from the Fourier transform of hxy and is normally referred to as the transfer function with magnitude and phase. The coherence squared function [8] can be used to gauge the quality of the system identification procedure. It is determined using Equation 2, where Sxy(jω), Sxx(jω) and Syy(jω) are the cross, stress and strain power spectral densities respectively. If there is any sensor noise or non linearity within the system the coherence squared will be less than one.

$$coh^2(j\omega) = \frac{\left|S_{xy}(j\omega)\right|^2}{\left|S_{xx}(j\omega)\right|\left|S_{yy}(j\omega)\right|} \qquad 0 \le coh^2(j\omega) \le 1 \qquad (3)$$

Mechanical Testing

An initial test was conducted to measure the compliance without any electrochemical stimulus. This test was used to calculate the compliance impulse response using the stress input described in the above section. The data was sampled at 1000Hz.

Electrochemical Mechanical Testing:

In this test two separate inputs are used to excite the polymer, an electrochemical potential and a stochastic stress. A three electrode electrochemical cell was coupled with a dynamic mechanical analyzer [7] to generate the inputs. A silver wire was used as a reference electrode and the tests were conducted in neat 1-butyl-3-methylimidazolium hexaflourophosphate (BMIMPF6). The samples were warmed up by cyclic voltametry until the electrochemical response was stabilized. The voltage input was a triangle wave between -0.8 V to 0.8 V and a rate of 20 mV/s. The stress input was the same as described in the mechanical testing section above. The resulting strain was a high frequency strain superimposed on a low frequency strain (Figure 3). The stress and strain data were filtered using a 2nd order low pass butterworth anti aliasing filter implemented in hardware with a cut off frequency at 100Hz.

RESULTS AND DISCUSSION

Mechanical Characterization

Figure 1 shows the calculated compliance impulse response of polypyrrole. The impulse response rises to 200 GPa^{-1} and falls to zero within 20 lags or 0.02s. This indicates that the 40 point impulse response was sufficient to fit the data.

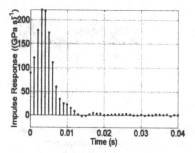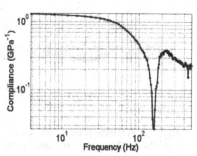

Figure 1: Left: The calculated compliance impulse response of polypyrrole. Right: The compliance frequency response of polypyrrole calculated using Gaussian stress strain data and Equation 1.

Figure 1 also shows the frequency domain equivalent of compliance impulse response also known as the transfer function. This indicates that the system is at least 3rd order. The low frequency compliance was 2 GPa^{-1} which corresponds to an elastic modulus of 500 MPa. This matches with prior studies conducted on the same material [1]. There was a gradual reduction in the compliance until 100 Hz, after which there was a large drop in the compliance.

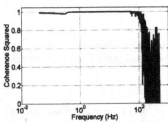

Figure 2: Coherence squared of the stress strain data. This indicates that a linear model is sufficient to model the compliance data up to 100Hz.

However, the coherence squared (Figure 2) estimate was close to one only until 100 Hz after which there was a large drop. This would indicate that the compliance response shown in Figure 1 was only valid up to 100 Hz. Since the coherence was close to one until 100 Hz, a linear model represented by the compliance impulse response was sufficient to model the data. Our input stress is within a range of 5-10MPa and it should be noted that the response calculated is accurate within that stress range.

Electrochemical Characterization

Ionic liquids such as BMIMPF6 can be used as solvents for actuation. In neat BMIMPF6 the BMIM$^+$ cation tends to diffuse into the polymer over a wide range of potentials [1, 9-10]. Therefore the expansion of the polypyrrole occurs during reduction when the cation diffuses into the film. We do not use any solvent to avoid any changes to the compliance due to solvent transfer [1].

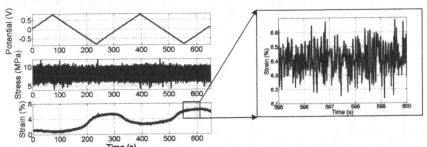

Figure 3: Polypyrrole being excited using a triangle wave voltage input that varies at 10mV/sec and the stochastic stress input. The output strain has a low frequency volumetric component proportional to the charge superimposed on a high frequency strain.

Figure 3 shows polypyrrole being excited by 2 inputs, a high frequency stochastic stress wave and a low frequency voltage waveform. The resulting strain was composed of a low frequency strain proportional to the charge injected into the polymer and a high frequency strain due to input stress. The resulting stress and strain data was sectioned into sequences of 15 seconds and the impulse response and frequency response for each of the sections was calculated. Since the frequency of the applied voltage was low, there was little change to the compliance during this 15 s interval.

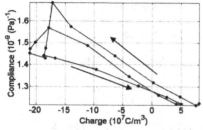

Figure 4: The low frequency compliance of the polypyrrole as a function of charge injected into the polymer.

The compliance impulse responses are strongly correlated with the charge injected into the polymer. The low frequency gain of the transfer function shows the most significant change. Figure 4 shows the changing low frequency compliance due to charge injection. As the $BMIM^+$ diffuse into the polypyrrole the low frequency compliance changes significantly by 40-50% during each cycle. The $BMIM^+$ seems to only change the low frequency component of the compliance transfer function and does not affect the natural frequency or the damping of the compliance.

CONCLUSIONS

We have used a novel technique to measure the compliance transfer function of polymers. Based on the coherence squared estimate the compliance transfer function was found to be valid up to 100 Hz. We have also shown that the compliance impulse response function changes as $BMIM^+$ ions are driven in and out of the polymer. The low frequency polymer compliance changes by as much as 40-50% as ions diffuse through the polymer. There is no discernable effect of the ions on the natural frequency or the damping of compliance transfer function. Future tests will include the use of solvents and other ions to see how they affect other features of the transfer function.

ACKNOWLEDGMENTS

The authors would like to thank the National Science Foundation for providing a graduate fellowship to support this work. Research is supported by the Institute of Soldier Nanotechnologies supported by the US Army research Laboratories and the US Army research office under Contract No DAAD-19-02-002.

REFERENCES

1. Pytel R.Z., Thomas E.L, and Hunter I. W., Polymer. **49**, 2008-2013, (2008).

2. Y.B. Cohen, "Electroactive Polymer [EAP] Actuators as Artificial Muscles- reality, potential and challenges" ed Y. B Cohen. SPIE press (2001)
3. Spinks GM, Liu L, Wallace GG, and Zhou DZ. Advanced Functional Materials **12** (6-7):437-440 (2002)
4. Bay L, Mogensen N, Skaarup S, Sommer-Larsen P, Jorgensen M, and West K. Macromolecules, **35**(25):9345-9351 (2002).
5. Otero TF, Lopez Cascales JJ, and Vazquez Arenas G. Materials Science and Engineering: C **27**(1):18-22 (2007)
6. Samani MB, Cook DC, Madden JD, Spinks G, and Whitten P. Thin Solid Films, **516**:2800-2807 (2007)
7. Vandesteeg N. PhD. Thesis. Synthesis and Characterization of Conducting Polymer Actuators. Materials Science and Engineering, Massachusetts Institute of Technology, (2007)
8. Jer Nan Juang. "Applied System Identification". Englewood Cliffs, NJ: Prentice Hall; 1994.
9. Vandesteeg N, Anquetil P, and Hunter I. Poly(3,4-ethylenedioxythiophene) actuators: the role of cation and anion choice. SPIE Smart Structures and Materials 2004: Electroactive Polymers Actuators and Devices. San Diego, CA, 2004
10. Ding J, Zhou DZ, Spinks G, Wallace G, Forsyth S, Forsyth M, and MacFarlane D. Chemistry of Materials,15(12):2392-2398 (2003)
11. Pieter Eykhoff. "System Identification: Parameter and State Estimation". John Wiley & Sons; (1974)

Mater. Res. Soc. Symp. Proc. Vol. 1134 © 2009 Materials Research Society   1134-BB07-06

Elastic Aerogels and Xerogels Synthesized From Methyltrimethoxysilane (MTMS)

Kazuyoshi Kanamori,* Kazuki Nakanishi and Teiichi Hanada
Department of Chemistry, Graduate School of Science, Kyoto University,
Kitashirakawa, Sakyo-ku, Kyoto 606-8502, Japan
*Corresponding author (e-mail: kanamori@kuchem.kyoto-u.ac.jp)

ABSTRACT

Transparent organic-inorganic hybrid aerogels and aerogel-like xerogels have been prepared from methyltrimethoxysilane (MTMS) respectively by supercritical drying (SCD) and ambient pressure drying (APD). The new aerogels and xerogels significantly deform without collapsing on uniaxial compression and almost fully relax when unloaded. This elastic behavior, termed as "spring-back", allows APD without noticeable shrinkage and cracking. The flexible network composed of lower cross-linking density (up to three bonds per every silicon atom) compared to silica gels (up to four bonds) and repulsion between hydrophobic methyl groups bonded to every silicon atom largely contributes to the pronounced deformability and relaxing, respectively. Lower surface silanol group density also plays a crucial role for the "spring-back" behavior.

INTRODUCTION

Silica aerogels are generally prepared by supercritical drying (SCD) of a wet gel obtained from a liquid-phase synthesis such as the sol-gel method [1,2]. Owing to the high porosity (> 90 %) and small pore size (~ 50 nm), silica (SiO_2) aerogels show outstanding properties such as high transparency, low refractive index, extremely low thermal conductivity and low dielectric constant. Silica aerogels thus have been intensively studied for decades, however, the complicated production process including SCD and substantial fragility have been preventing aerogels from extended applications. The low mechanical properties are attributed from the fact that diminutive silica particles (typically ~ 10 nm) are point-contacted each other to form the dilute solid phase. If the seriously fragile property is improved, both production and handling of aerogels become much easier.

Silica aerogels are usually prepared from water glass or silicon alkoxide such as tetraalkoxysilane ($Si(OR)_4$), which is a typical precursor used in the sol-gel method. For the purpose of improving the mechanical properties, we employed methyltrimethoxysilane (MTMS) as a single precursor in the sol-gel method to obtain organic-inorganic hybrid aerogels and xerogels (dried under ambient pressure) [3,4]. Although using MTMS as a single precursor results in macroscopic phase separation and inhomogeneous gelation due to cyclization in most cases [5], here we report a successful preparation of transparent monolithic polymethylsilsesquioxane (PMSSQ, $MeSiO_{1.5}$) aerogels with improved mechanical properties by effectively suppressing macroscopic phase separation and cyclization. Some of the obtained PMSSQ aerogels show significant deformation against uniaxial compression and the perfect recovery when unloaded (so-called "spring-back"). Owing to this elastic behavior, also obtained are aerogel-like xerogels from ambient pressure drying (APD).

EXPERIMENTAL

Acetic acid, distilled water, urea, and 2-propanol were purchased from Hayashi Pure Chemical Ind., Ltd. (Japan). Surfactant n-hexadecyltrimethylammonium bromide (CTAB) and poly(ethylene oxide)-$block$-poly(propylene oxide)-$block$-poly(ethylene oxide) ($EO_{106}PO_{70}EO_{106}$, F127) were from Tokyo Chemical Ind. Co., Ltd. (Japan) and BASF SE (Germany), respectively. Methyltrimethoxysilane (MTMS) was obtained from Shin-Etsu Chemical Co., Ltd. (Japan). Methylnonafluorobutylether (Novec HFE-7100) was purchased from Sumitomo 3M Ltd. (Japan). All the reagents were used as received.

The starting compositions are listed in Table 1. Given amounts of aqueous acetic acid (1, 5 or 10 mM), surfactant (F127 or CTAB) and urea were dissolved in a glass sample tube, and then MTMS was added with vigorous stirring. The mixed solution was stirred for 30 min at room temperature for hydrolysis, and followed by gelation at 60 °C in a closed vessel. Urea hydrolyzes into ammonia and carbon dioxide to raise pH of the sol at 60 °C, which, a modified 2-step acid/base process, allows random cross-linking rather than cyclization. The wet gel was then aged for 2 d to complete the condensation, followed by washing with distilled water for 24 h at 60 °C. For SCD, the washed sample was subjected to a solvent-exchange with 2-propanol at 60 °C for 24 h for 3 times. Alcogels obtained in this way were dried from supercritical carbon dioxide at 80 °C, 13.5 MPa in a custom-built autoclave (Kobe Steel Ltd., Japan). For APD by evaporation, the washed sample was solvent-exchanged with methanol at 60 °C for 8 h at reflux for 3 times, followed by a solvent-exchange with methylnonafluorobutylether at 55 °C for 8 h at reflux for 3 times. They were then slowly evaporative-dried at 30 °C for 3 d to obtain xerogels.

To observe the pore morphology, FE-SEM (JSM-6700F, JEOL Ltd., Japan) was employed. Bulk density, ρ_b, was obtained by weighing the sample of known volume. Porosity ε was then determined as $\varepsilon = (1 - \rho_b/\rho_s)$, where ρ_s represents skeletal density that was fixed to be 1.35 g·cm^{-3} determined for a dried MTMS-derived PMMSQ gel. For light transmittance measurements, a UV/Vis spectrometer (V-570, JASCO Corp., Japan) equipped with an integrating sphere (ISN-470) was employed. The obtained total transmittance data at 550 nm were normalized into those of 10 mm-thick samples using the Lambert-Beer equation, and denoted as T.

Table 1   Starting compositions and obtained properties of aerogels and xerogels

	acetic acid aq.	surfactant	MTMS /g	urea /g	ρ_b /g cm^{-3}	ε	T
MF10-100a	10 mM, 10.00 g	F127, 1.00 g	4.76	0.50	0.18	0.867	0.159
MF5-100a	5 mM, 10.00 g	F127, 1.00 g	4.76	0.50	0.18	0.867	0.212
MF1-100a	1 mM, 10.00 g	F127, 1.00 g	4.76	0.50	0.19	0.859	0.239
MC10-100a	10 mM, 10.00 g	CTAB, 1.00 g	4.76	0.50	0.20	0.852	0.358
MC5-100a	5 mM, 10.00 g	CTAB, 1.00 g	4.76	0.50	0.22	0.837	0.617
MC1-100a	1 mM, 10.00 g	CTAB, 1.00 g	4.76	0.50	0.21	0.844	0.574
MF10-110a	10 mM, 7.00 g	F127, 1.10 g	4.76	0.50	0.22	0.837	0.523
MF10-110x	10 mM, 7.00 g	F127, 1.10 g	4.76	0.50	0.23	0.830	0.462
MC10-060a	10 mM, 7.00 g	CTAB, 0.60 g	4.76	0.50	0.21	0.844	0.366
MC10-060x	10 mM, 7.00 g	CTAB, 0.60 g	4.76	0.50	0.24	0.822	0.251

RESULTS & DISCUSSION

Aerogels Obtained from SCD

Table 1 and Figure 1 exhibit the starting compositions and resultant aerogel morphologies obtained by SCD. For the sample notation, gels obtained from F127 and CTAB systems start with "MF" and "MC", respectively. The following number denotes the concentration of acetic acid in mM, and a number after the hyphen is related with the weight of surfactant. The last character "a" or "x" represents SCD aerogel or APD xerogel, respectively. From Figure 1a–c, it can be confirmed that aerogels obtained from MF system obviously consist of fibrous skeletons and mainly large pores (~100 nm, macropores). Whereas aerogels obtained from CTAB system possess skeletons of aggregated globules and smaller pores (~30 nm, mesopores) as found in Fig. 1d–f. Since well-developed microphase-separated gels are obtained if no surfactant is added, the role of surfactant here is considered to be suppressing phase separation. The morphology of aerogels obtained from MF system is therefore regarded as the transient bicontinuous structure frozen in the very early stage of spinodal decomposition. Although the phase fraction is significantly asymmetric as can be imagined by high porosity, aerogels from MF system retain bicontinuous structure because of the high viscosity of the sol. In MC system, on the other hand, the morphology is much more like conventional silica aerogels, and there is no trace of phase separation. This suggests that CTAB solubilizes hydrophobic MTMS oligomers/polymers more effectively in the aqueous solution.

Changing the concentration of acetic acid somewhat affects the morphologies of aerogels. For MF system, domain size (pore size plus skeleton size) is hardly changed by varying the concentration of acetic acid, which can be corroborated by comparing the FE-SEM images of MF10-100a, MF5-100a and MF1-100a (parts a–c). For MC system, the effect of acid concentration is much evident. In Fig. 1d–f, it can be confirmed that the pore morphology becomes more homogeneous with decreasing concentration of acetic acid. For example, MC10-100a in Fig. 1d, compared to MC5-100a (part e) or MC1-100a (part f), exhibits agglomerated parts as circled in Fig. 1d. Since similar behavior can be seen with increasing amount of urea while fixing the concentration of acetic acid, this behavior is attributed to the progress of hydrolysis and to the pH change during condensation of MTMS. In addition to the fact that hydrolysis of MTMS becomes slower due to the stronger solubilization by CTAB, less concentration of acetic acid results in more incomplete hydrolysis of MTMS. The less hydrolyzed MTMS forms fewer number of linkage, i.e. condensation is more or less inhibited, and smaller particles result [6]. Since less concentration of acetic acid results in higher pH during condensation, the surface charge density of MTMS oligomers becomes more negative. Preferential aggregation of colloids is therefore avoided in the higher pH due to the more electrostatic repulsion in-between the particles. In this manner, smaller particles are more homogeneously aggregated in the samples derived from more dilute acetic acid such as MC5-100a and MC1-100a, and consequently, higher light transmittance is achieved. Conversely, increasing concentration of acetic acid leads to more complete hydrolysis and less electrostatic repulsion, resulting in much aggregation of colloidal particles. Accordingly, inhomogeneous porous network forms, which leads to lower light transmittance as can be seen in MC10-100a.

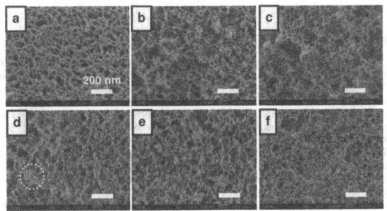

Figure 1   FE-SEM images of obtained aerogels; (a) MF10-100a, (b) MF5-100a, (c) MF1-100a, (d) MC10-100a, (e) MC5-100a, and (f) MC1-100a. The circled portion in part d represents the highly aggregated particles which cannot be seen in (e) and (f).

Mechanical Properties of Aerogel

The PMMSQ aerogels prepared purely from MTMS show a drastic improvement of mechanical durability against compression. Figure 2 shows a uniaxial compression test on the sample MC10-060a. The gel shows up to 80 % strain and spring-backs to the original size when unloaded. The significant deformation like shown here has not been observed for pure silica aerogels. Three factors specific for PMMSQ networks are attributed to this phenomenon. First, the fact that only up to three siloxane bonds per each silicon atom lowers the crosslinking density and makes the gel more flexible than pure silica which is composed of up to four siloxane bonds. Second, low density of silanol groups [7] hinders the nonreversible shrinkage. In pure silica, the permanent shrinkage is usually observed because silanol groups with high density form additional siloxane bonds when the gel is forced to shrink during drying [8], whereas this is avoided in PMMSQ. Third, methyl groups are homogeneously distributed with high density in the network, which will repel each other during the temporal shrinkage on compression. These three factors contribute to the significant deformation, avoiding nonreversible shrinkage, and the spring-back behavior, respectively.

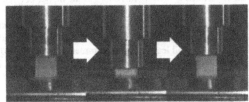

Figure 2   Overview of a uniaxial compression test on MC10-060a showing a drastic linear strain as large as 80% and subsequent perfect spring-back.

Possibility of APD

The deformability against compression makes it possible to dry under ambient pressure with maintaining the original pore structure. During evaporative drying, considerable capillary pressure is exerted on the external surface, and thereafter inside the gel networks. According to Scherer [9,10], a drying gel starts shrinking by the capillary pressure originated from menisci on the outer surface during the early stage of drying, and stops shrinking at the critical point. Then, the menisci will intrude into the gel body, so the possibility of cracking increases. In the case of the present PMMSQ gels, the drying gel with starting composition MC10-060 also showed the similar behavior as in the compression test during APD. Namely, the gel drastically shrank (64 % in volume) by the capillary pressure from the external surface and then relaxed on emptying the pores. The obtained xerogel showed no macroscopic cracking and shrinkage. Gels from MF system also showed the similar or even better behaviors. A wet gel from MF10-110 composition, which has more continuous gel skeletons than MC10-060, was also dried under ambient conditions without causing a considerable structural change. The nanostructure of these gels shows some differences as presented in Fig. 3. In MF10-110 shown in parts a and b, only slight difference on the pore structures can be seen between the aerogel and xerogel from the FE-SEM images. On the other hand, the xerogel MC10-060x shown in part d caused noticeable deformation upon evaporative drying. By comparing with the aerogel MC10-060a in part c, it can be noticed that MC10-060x contains aggregated parts while in other parts there are larger pores. This obviously demonstrates that the particles in the original gel stick each other when forced to come close during the temporal shrinkage. This would result in the formation of smaller pores while leaving larger pores on the other. Somewhat disturbed pore structures cause higher light scattering, resulting in more significant reduction in T compared to MF10-110a and -x as listed in Table 1.

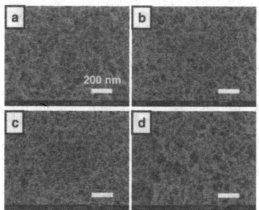

Figure 3   FE-SEM images of the aerogels and xerogels. (a) MF10-110a, (b) MF10-110x, (c) MC10-060a, and (d) MC10-060x.

CONCLUSIONS

Organic–inorganic hybrid polymethylsilsesquioxane (PMSSQ) aerogels and xerogels are obtained from methyltrimethoxysilane (MTMS) by utilizing a modified 2-step acid/base sol–gel route where urea is employed for base-releasing agent. To suppress the strong tendency for MTMS to phase-separate in the course of polycondensation, appropriate surfactant is employed, and it is found that n-hexadecyltrimethylammonium bromide (CTAB) and poly(ethylene oxide)-*block*-poly(propylene oxide)-*block*-poly(ethylene oxide) ($EO_{106}PO_{70}EO_{106}$, F127) are effective for this purpose. Aerogels from F127-containig system show larger pores with fibrous networks whereas those from CTAB-containing system are composed of networks made of aggregated colloidal particles and much finer pores.

It is found that PMSSQ aerogels show high mechanical durability against uniaxial compression, which gives us the possibility to dry a wet gel by the conventional evaporative drying process. Very recently, we have successfully obtained crack-free xerogels with light transmittance 0.85 as a $100 \times 100 \times 10$ mm^3 plate with typical bulk density of 0.10 g·cm^{-3} (porosity ~ 0.93) by APD. Details will be reported elsewhere.

ACKNOWLEDGMENTS

The present work was partly supported by a Grant for Practical Application of University R&D Results under the Matching Fund Method from New Energy and Industrial Technology Development Organization (NEDO), Japan. This research was also supported by the Global COE Program "International Center for Integrated Research and Advanced Education in Materials Science" (No. B-09) of the Ministry of Education, Culture, Sports, Science and Technology (MEXT) of Japan, administrated by the Japan Society for the Promotion of Science (JSPS). Also, this study was supported by Grant-in-Aid for Young Scientists (B) (No. 20750177) and by Scientific Research (B) (No. 20350094) from the MEXT, Japan.

REFERENCES

1. N. Hüsing, U. Schubert, *Angew. Chem. Int. Ed.* **37**, 22 (1998).
2. A. C. Pierre, G. M. Pajonk, *Chem. Rev.* **102**, 4243 (2002).
3. K. Kanamori, M. Aizawa, K. Nakanishi, T. Hanada, *Adv. Mater.* **19**, 1589 (2007).
4. K. Kanamori, M. Aizawa, K. Nakanishi, T. Hanada, *J. Sol-Gel Sci. Tech.* **48**, 172 (2008).
5. D. A. Loy, B. M. Baugher, C. R. Baugher, D. A. Schneider, K. Rahimian, *Chem. Mater.* **12**, 3624 (2000).
6. C. J. Brinker, K. D. Keefer, D. W. Schaefer, C. S. Ashley, *J. Non-Cryst. Solids* **48**, 47 (1982).
7. C. D. Volpe, S. D. E. Pagani, *J. Non-Cryst. Solids* **209**, 51 (1997).
8. C. J. Brinker, G. W. Scherer in *Sol-gel science: The physics and chemistry of sol-gel processing*, (Academic Press, 1990).
9. G. W. Scherer, *J. Non-Cryst. Solids* **100**, 77 (1988).
10. G. W. Scherer, *J. Am. Ceram. Soc.* **73**, 3 (1990).

Mater. Res. Soc. Symp. Proc. Vol. 1134 © 2009 Materials Research Society   1134-BB07-07

Aromatic Polyurea for High Temperature High Energy Density Capacitors

Yong Wang[1], Xin Zhou[1], Minren Lin[2], Sheng-Guo Lu[2], Junhong Lin[3], and Q. M. Zhang[1,2]

[1]Department of Electrical Engineering, The Pennsylvania State University,
[2]Materials Research Institute, The Pennsylvania State University,
[3]Materials Science and Engineering, The Pennsylvania State University,
University Park, PA 16802

ABSTRACT

We investigate aromatic polyureas which can be fabricated in the form of thin films through CVD. It was found that the polymer possesses a flat dielectric response (k~ 4.2 and loss <1%)) to more than 200°C. The frequency-independent dielectric properties in the investigated frequency range(1kHz~1MHz), low conductance, low dissipation factor (~0.005), high breakdown strength (>800MV/m), high energy density (>12J/cm^3) and high efficiency suggest this polymer can be a good candidate material for high temperature energy storage capacitors. Breakdown strength was analyzed with Weibull model over a broad temperature range (25°C ~180°C). Experimental results indicate that aromatic polyurea is more like a nonpolar linear dielectric material because of its highly cross-linked structures. The experiment results further show that this polymer maintains its high performance even at high temperatures.

INTRODUCTION

Capacitors with high electric energy density are highly desirable for a broad range of modern power electronic systems such as medical devices, hybrid electric vehicles (HEV), and power weapon systems [1]. Due to their high breakdown field and graceful failure, which lead to high device reliability, dielectric polymers are presently the primary choice for high energy density capacitors. Recently, very high energy density (>10J/cm^3) have been achieved in polyvinylidene fluoride (PVDF) based copolymers and terolymers [2]. However, all these high energy density polymers can only work under temperatures below 125°C. They are not suitable for harsh environment applications such as HEVs, power electric weapons and aerospace systems which require operating temperature at least 150°C [3]. For high temperature polymer capacitors, polyimide is currently the most commonly used polymer material. Two other high temperature polymers Teflon perfluoroalkoxy (PFA) and poly-P-xylene (PPX), were also studied by some researchers for high temperature capacitors [4]. But all these high temperature materials do not have high energy density because of their low dielectric constants and low breakdown fields (<300MV/m) [4]. Therefore, high energy density polymers which can be operated under high temperatures will be very attractive for high power capacitor applications in harsh environments.

In this paper, we investigated an aromatic polyurea dielectric film for possible high temperature high energy density capacitor applications. Aromatic polyureas have been investigated for piezoelectric and non-linear optic applications for more than a decade [5]. These studies have revealed that the polymer also possesses a high thermal stability to above 200°C.

The aromatic polyurea films can be easily fabricated by vapor phase deposition which makes it possible to produce high purity dielectric films, leading to high electric breakdown field and high energy density. Its relatively high dielectric constant (K ~ 4.2) is attractive since the energy density U is directly proportional to the dielectric constant,

$$U = 1/2 \ \varepsilon_0 K E_m^2 \qquad (1)$$

where ε_0 is the vacuum permittivity (=8.85x10^{-12} F/m) and E_m is the maximum electric field.

SAMPLE PREPARATION AND EXPERIMENT

The aromatic polyurea P(MDA/MDI) thin films investigated were fabricated using a vacuum thermal vapor deposition process with two monomers: 4,4'-diaminodiphenylmethane (MDA) and 4,4'-diphenylmethane diisocyanate (MDI). The evaporation temperatures were 100°C for MDA and 75°C for MDI, respectively. The temperature of the substrate was 20°C. The precise control of the evaporation temperatures and substrate temperature are crucial in controlling the film stoichiometry. The pressure inside the vacuum chamber was about 5 x 10^{-6} torr. The deposition rate was 5Å/s. The as-deposited films were then annealed at 200°C for 30min to achieve full polymerization among oligomers produced during the vapor deposition process. The typical film thickness in this investigation was 2.5 μm and evaporated Al electrodes of 1 mm in diameter were used for the electric characterization. The dielectric properties were characterized at different frequencies and temperatures using an HP 4284A LCR meter equipped with a computer controlled temperature chamber. The polarization characterization was carried out with a Sawyer-Tower circuit from which the discharged energy density was determined. The frequency of the driving voltage was 1 kHz. The thermo-gravimetric analysis (TGA) (TGA 2050) was used to further characterize the thermal stability of the polymer films up to 500°C.

CHARACTERIZATION RESULTS AND DISCUSSION

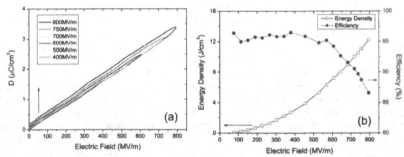

Figure 1. (a) The electric displacement versus electric field measured under 1 kHz for different applied field levels. (b) Discharged energy density (open squares) and charge-discharge efficiency (solid circles) versus electric field deduced from the data in (a).

Presented in Figure 1(a) is the charge-discharge displacement-electric field (D-E) curve measured at room temperature under different applied electric fields. The data show that the film can reach a breakdown field near 800 MV/m. The discharged (released) energy density, which is the integration $U_r= \int EdD$ from the discharge curve (upper branch of the D-E curve), is presented in figure 1(b). An U_r higher than 12 J/cm^3 is observed at 800 MV/m. It is also found that U_r fits well with eq. (1) with a dielectric constant K=4.2, indicating that the aromatic polyurea is a linear dielectric material.

The ratio between the U_r and U_{in}, where U_{in} is the total input energy density and is the integration $U_{in}= \int EdD$ from the charging curve (lower branch of the D-E curve), defines the efficiency of the dielectric material, also presented in Figure 1(b). The data shows that at fields below 550MV/m, the P(MDA/MDI) films display an efficiency of 95%, similar to those of the low loss polymer dielectrics such as biaxially oriented polypropylene(BOPP). At higher fields, the efficiency drops, due to high field tunneling current, which is common to all the dielectric materials [6].

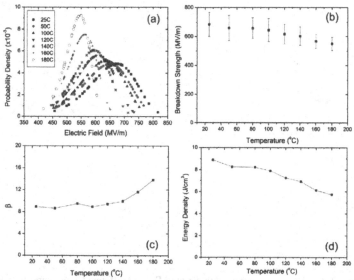

Figure 2. (a) Weibull distribution of the breakdown field of P(MDA/MDI) films at different temperatures, (b) Characteristic breakdown strength E_b as a function of temperature, (c) β as a function of temperature, deduced from the two parameter Weibull analysis. (d) Temperature dependence of energy density calculated from E_b in (b) based on the equation (1).

The electric breakdown strength of P(MDA/MDI) was characterized as a function of temperature. For each temperature, D-E curves were measured over 50 film samples and the data are presented in Figure 2(a). The data show that while the breakdown fields at the lower field end is more or less fixed near 450 MV/m, which is presumably caused by the defects in the films, the higher field end of the breakdown field decreases with temperature, indicating that the breakdown strength of the defect free films reduces with temperature. The two-parameter

Weibull analysis is used for breakdown analysis of the dielectric films [7],

$$P_f = 1 - \exp[-(E / E_b)^\beta] \qquad (2)$$

where E is the measured DC breakdown electric field, β is a shape parameter, and E_b is the characteristic breakdown strength (63.2% of accumulated probability of breakdown). A large β corresponds to a narrow distribution of the breakdown field. E_b and β thus obtained are summarized in figure 2(b) and 2(c). There is only a slight decrease of E_b with temperature and even at 180°C, E_b is still more than 500MV/m, attractive for high temperature applications. The discharged energy density U_r calculated from E_b in figure 2(b) based on equation (1) is presented in figure 2(d). Due to the reduction of E_b with temperature, U_r is also reduced from near 9 J/cm^3 at room temperature to about 6 J/cm^3 at 180°C.

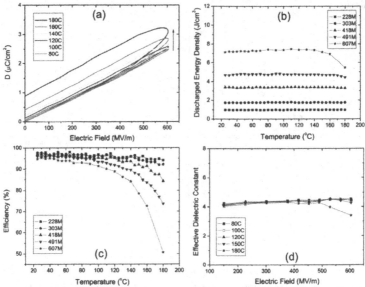

Figure 3. (a) D-E curves measured under 600MV/m at different temperatures (curves are plotted with increasing temperature from bottom to top). (b) Discharged energy density U_r versus temperature under different electric fields. (c) Charge-discharge efficiency versus temperature under different electric fields. (d) Field dependence of effective dielectric constant K_{eff} calculated from $U_r = 1/2\ K_{eff}\varepsilon_0 E^2$ at different temperatures.

Figure 3(a) presents the D-E curves at 600 MV/m at different temperatures. As has been pointed out in our previous study, all the dielectric materials will suffer tunneling current loss at high electric field. For the P(MDA/MDI) films here, this high field conduction loss (upper end in D-E curve when E=0) increases with temperature. The temperature dependences of discharged energy density and efficiency were measured under different fields and shown in figures 3(b) and 3(c). Under low fields (<500MV/m), the discharged energy density is nearly a constant under a given field up to 180°C. On the other hand, the efficiency shows large drop with temperature for fields above 400 MV/m as temperatures approach 180°C, caused by the increased high field conduction loss at high temperatures which shows exponential dependence on temperature. Such a loss behavior can be caused by the de-trapping of charge carriers (charge carrier hopping in the polymer films) such as the Poole-Frenkel effect which can become significant at high temperature and high field.

In many polymers, a material with high dielectric constant at low field does not mean it still possesses a high energy density due to the possible high field dielectric saturation. To describe such a dielectric saturation effect, an effective dielectric constant K_{eff} is used for the discharged energy density $U_r = 1/2 K_{eff}\varepsilon_0 E^2$ and a constant K_{eff} with field indicates that there is no high field dielectric saturation effect. K_{eff} for the P(MDA/MDI) films measured at different temperature as a function of applied field is presented in figure 3(d). The data show that K_{eff} is a constant for temperatures below 150°C and fields below 600 MV/m. The apparent reduction of K_{eff} at 180°C and above 500 MV/m is probably caused by the increased high field tunneling current and space charges which modify the effective field in the dielectric films.

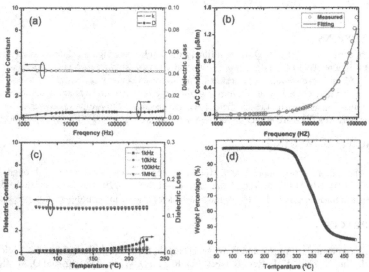

Figure 4. (a) Frequency-dependence of dielectric properties of P(MDA/MDI) films at room temperature. (b) Conductivity σ as a function of frequency, open circles are data points and the solid curve is a fitting of σ ~ ω. (c) Temperature-dependence of dielectric properties at different frequencies. (d) TGA data of the P(MDA/MDI) films.

We also studied the low field dielectric properties of P(MDA/MDI) films. Figure 4(a) presents the frequency dependence of the dielectric constant and loss measured from 1 kHz to 1MHz at room temperature. The dielectric constant is about 4.2 and dielectric loss tanδ is 0.005 at 1 kHz and room temperature. To further probe the frequency dependence of dielectric response and the charge transport mechanism, we carried out the analysis of the conductivity σ from the complex dielectric permittivity $\varepsilon(\omega)^* = \varepsilon' - j\sigma/\omega$, where ε' is the real part of the dielectric permittivity, ω is the angular frequency [8]. It is found that for $\sigma = \sigma_0 + \sigma'(\omega)$, where σ_0 is the DC (zero frequency conductivity) and $\sigma'(\omega)$ is the AC component of the conductivity, $\sigma'(\omega)$ dominates and σ_0 is nearly zero for all the films studied here (in thickness from 100nm to 8 μm) (see figure 4(b)). It is further found that $\sigma \sim \omega^n$ with n=1, suggesting that the hopping conduction process dominates the dielectric loss, which is also consistent with the high temperature high field loss data. Such a hopping process can be from the interchain electron transfer from amine groups (as donors) to urea groups (as acceptors).

Figure 4(c) presents the dielectric properties of P(MDA/MDI) as a function of temperature for different frequencies. The dielectric properties are nearly constant to temperatures up to 200°C. The increased dielectric loss above 200°C is due to the thermal enhanced conduction of charge carriers in the polymer. To further investigate its thermal behavior, TGA experiment was carried out. As can be seen in figure 4(d) there is no observable weight change to above 200°C, and the weight of the sample starts to drop slowly above 250°C which indicates the polymer starts to decompose. These data indicate that the polymer films studied here are suitable for applications to temperatures higher than 200°C.

CONCLUSIONS

In conclusion, P(DMA/DMI) thin films exhibit relatively high breakdown strength (>500 MV/m) with a relatively high dielectric constant to 180°C, yielding a high discharged energy density (>6J/cm^3) even at high temperatures. The increased conduction loss at 180°C and above 500MV/m is likely caused by the de-trapping of charge species and the hopping conduction, which is also consistent with the low field dielectric loss data. The results suggest that the aromatic polyurea films are attractive for high temperature high energy density capacitors.

ACKNOWLEDGMENTS

The authors greatly appreciate the financial support of this work by ONR through Grant No. N00014-05-1-0541 (MURI) and Grant No. N00014-08-1-0229. The authors also thank Dr. Takahashi in So-Ken Co., Ltd., Tokyo, Japan for his assistance in the polymer film preparation.

REFERENCES

1. W. J. Sarjeant, J. Zirnheld, and F. W. MacDougall, IEEE Trans. Plasma Sci. **26,** 1368 (1998).
2. B. J. Chu, X. Zhou, K. L. Ren, B. Neese, M. R. Lin, Q. Wang, F. Bauer, and Q. M. Zhang, Science **313,** 334 (2006).
3. R. Kirschman, *High-Temperature Electronics* (Wiley-IEEE Press, 1998).
4. W. Khachen, J. Suthar, A. Stokes, R. Dollinger, IEEE Trans. Electr. Insul. **28,** 876 (1993).
5. Y. Takahashi, S. Ukishima, M. Iijima, and E. Fukada, J. Appl. Phys. **70,** 6983 (1991).
6. Q. Chen, Y. Wang, X. Zhou, Q. M. Zhang, Appl. Phys. Lett. **92** 142909 (2008).
7. L. A. Dissado and J. C. Fothergill, *Electrical Degradation and Breakdown in Polymers* (P. Peregrinus, London, 1992).
8. M. O. Aboelfotoh and C. Feger, Phys. Rev. B **47,** 13395 (1993).

Poster Session

Foster Session

Mater. Res. Soc. Symp. Proc. Vol. 1134 © 2009 Materials Research Society 1134-BB08-03

Direction sensitive deformation sensing with CNT/epoxy nanocomposites

S.T. Buschhorn, M.H.G. Wichmann, J. Gehrmann, L. Böger, K. Schulte
Technische Universität Hamburg-Harburg, Institute of Polymers and Composites, Denickestr. 15, D-21073 Hamburg, Germany

ABSTRACT

In this work direction sensitive bending strain sensors consisting of a single block of epoxy/multi-wall carbon nanotube composite were developed. The nanocomposite electrical resistance correlates closely with mechanical strain. The directional sensitivity to bending deformations is related to the change in electrical resistance, which becomes positive or negative, depending on the direction of bending deflection. This effect is brought about due to an inhomogenous distribution of electrical conductivity throughout the material. The production process was performed in a straightforward single-step processing route. In a modified setup the resistance change of the nanocomposite is related to exclusively compressive loads or tensile loads. A characterisation of the conductivity distribution within the produced material was done in order to find the driving forces behind the phenomenon.

INTRODUCTION

Carbon nanotubes are known to be suited candidates for the implication of electrical conductivity and sensing capabilities [1] in polymer matrix systems. In epoxy matrix systems especially low percolation thresholds can be achieved [2]. During the curing process of the resulting nanocomposite, the carbon nanotubes can organize in a way which facilitates an electrically conductive network. If the nanocomposite is strained mechanically the resulting changes in the conductive network structure will be reflected in a variation of the nanocomposite bulk conductivity.

This work concentrates on the interrelation between production process of bulk samples and the resulting piezoresistive behavior of epoxy based carbon nanotube composites. Bending load was applied in a three point bending beam setup while the resistance response of the nanocomposites was recorded simultaneously.

For CNT/epoxy nanocomposites the resistivity was found to correlate closely with mechanical strain. Under tensile stress, for low strain values below 2.5%, the electrical resistivity was found to increase in a nearly way with the applied strain [3, 4]. For higher strain values the resistance curve deviates significantly from the initial behavior as non elastic deformation mechanisms local deformation of the epoxy matrix become more dominant.

Under bending deformation, the nanocomposite systems revealed a pronounced direction dependent sensitivity. It was found that the curing process effects the formation of the conductive network significantly. A slight inhomogeneity in the bulk sample was sufficient to introduce significant electromechanical anisotropy. Thus, direction sensitive bending sensors could be made with relatively simple methods.

EXPERIMENT

Materials and sample preparation

The samples were produced using a hot curing epoxy system (Araldite LY 556, Aradur 917, and Accelerator DY 070, made by Huntsman). The initial high viscosity epoxy resin facilitates a fine dispersion of carbon nanotubes (Graphistrength C100, Arkema) in the resin.

The dispersion is attained through the use of a high shear force three roll mill (120E, EXAKT). The mill's rolls rotate at different speeds and are oriented with their axis of rotation in parallel and spaced so that the rounded surfaces of two adjacent rolls form a small gap of adjustable width. A mixture of the epoxy resin and multiwall carbon nanotubes is poured onto the roll and is sheared in the gaps between the rotating rolls. Different methods have been used to attain the desired dispersion. However, they produce comparable results and vary mostly is material throughput.

One method can be characterized through a dwell time of 2 minutes for which a small batch remained on the rolls while they were rotating at 30 rpm, 90 rpm, and 180 rpm. After removal another batch was applied to the rolls. The other method can be characterized through the rotational speed, which was increased to 300 rpm. And the number of times a single large batch of the mixture passed through such gaps, which was 16 in total. The gap size varied in the beginning of the second process in order to effectively break apart the initially large agglomerates. However majority of the passes were through gaps with a width of 5 μm, thus the mixture was subjected to a shear rate in excess of $1.4*10^5 \, s^{-1}$ several times.

The resin was subsequently mixed with the low viscosity anhydride hardener and accelerator in a ratio of 100 : 90 : 1 parts by weight respectively. The resulting mixture then contained 0.1 wt. % MWCNTs and was cast into a thick walled, open topped aluminum mold, and cured in a panel heated oven. The curing cycle was set to 4 h at 80 °C and 8 h at 140 °C. In order to find out more about the influence of the curing process another batch was cured in a wall heated oven set to the same curing cycle.

Two kinds of samples were produced. One consisted of a single layer of nanocomposite cut into beams of approximately 7 mm thickness, 10 mm width, and 100 mm length. The other type was made out of two layers of reduced thickness, of which one layer consisted of neat epoxy cured on top of the nanocomposite in an additional step. These were produced so the interface between the layers would coincide with the neutral fiber in a bent sample. The samples were contacted with a wire, which was held in place by conductive silver paint. A schematic of the production process is given in Figure [1].

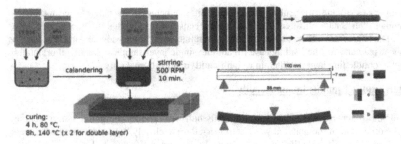

Figure 1: Schematic illustration of sample production and the bending beam setup with the different loading cases a, b, c for double layers samples and single layer samples.

Characterization of piezoresistive behavior

The three point bending beam setup used a support distance of 88 mm and the results were obtained for flexures not exceeding 2 % strain of the outer fiber as. Electrical measurements were performed using a Keithley 2602 sourcemeter in a 2-electrode DC setup using a potential difference of 10 V with a data acquisition rate of 5 at least Hz. Preliminary experiments with a four-point setup showed that the influence of the contact resistances was negligible. Several measurements were performed for the different kinds of beams in order to elucidate the influence flexture has on its resistance. Figure [2] shows a exemplary diagram. More details concerning the strain resistance relationship are available in [4].

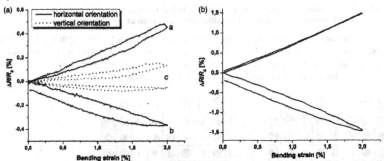

Fig. 2(a): Relationship between strain of the outer fiber and the resistance change for different loading cases in a single layer beam: a – high conductivity under tensile strain, b – high conductivity under compressive strain, c – highly conductive side parallel to load direction
Fig. 2(b): Relationship between strain of the outer fiber and the resistance change for different loading cases in double layer beams: resistance increase when conducting half is under tensile strain, resistance decrease when conductive half is under compressive strain.
The samples were produced using the panel heated oven.

The loading cases (a) and (b) result in the same qualitative behavior for single layer and double layer samples. The main difference is the higher sensitivity exhibited by the double layer samples. The observed effect is essentially reversible, yet some degree of hysteresis is observable, especially for the loading case (b) and the single layer samples. A vertical orientation of the highly conducting layer resulted in a significantly reduced sensitivity.

Characterization of sample inhomogeneity

Of each plate of nanocomposite that was cured only half was used for the production of bending beams. The other half of the material was characterized more closely. It was cut into three layers, which in turn were cut into tiles of approximately 10 mm x 10 mm x 1 mm. These were contacted with conductive silver paste on their large face. In this way it was possible to characterize the conductivity of the produced material with at different locations. Samples were produced for three height levels and a raster of 9 by 7 samples. The results for the electrical conductivity of the three height levels as well as the types of oven can be seen in Figure [3].

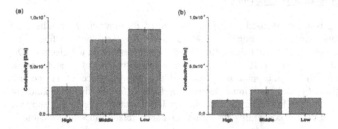

Figure 3(a): Electrical conductivities measured for different height levels in a panel heated oven.
Figure 3(b): Electrical conductivities measured for different height levels in a wall heated oven.

The electrical conductivity within one height level was also found to encompass a significant range. Between the different kinds of ovens a qualitative difference was visible as illustrated in Figure [4]. The distribution of conductivities for the different height levels (not shown) were less clear cut and would need more samples in order to yield more robust results.

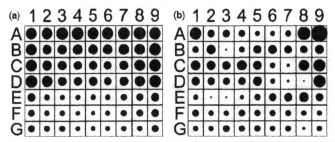

Figure 4: The area of a black circle is proportional to the relative conductivity of one 10 mm x 10 mm sample in the high layer of (a): a sample produced in a panel heated oven, (b): a sample produced in a wall heated oven.

The produced plates were cut along the line G1-G9, so that G5 would be close to the center of the plate. It should be noted that the absolute conductivities vary significantly between (a) and (b). It is obvious that the conductivity is generally higher at locations close to boundaries, which can be seen in both types of samples. The sample cured in the panel heated (a) oven shows a quite systematic and symmetrical distribution, whereas (b) only has high conductivity in the corners and some seemingly random variation across the rest of the area. It is interesting to note that local variations have characteristic sizes which allow them to be detected even with the coarse resolution of the sampling raster.

DISCUSSION

The direction sensitivity of the beams is a result of the inhomogeneous conducting network within the sample. When the sample is bent one side will be subjected to the maximum of tensile strain, while the opposite is subjected to the maximum compressive strain. Between them will be the maximum of shear force as can be described by the Euler-Bernoulli beam equation or Timoshenko beam theory [5]. For a linear elastic material that is homogeneous and isotropic in a pure bending the strain can be described as follows. At a certain distance y from the neutral surface it is $\varepsilon_x(y) = -\varepsilon_m y / c$, where ε_x is the strain along the length of the beam, ε_m is the maximum strain at the corresponding extreme surface, dependent of the radius of curvature, which is at a distance c from the neutral surface. An integration over the resistance change induced throughout two halves of the deformed sample is consistent with the measured results for single and double layer the bulk samples.

If the electric conductivity of a material is distributed symmetrically around the neutral axis, and the resistance change is linearly dependent on strain, the resistance change throughout the material will cancel out. As the resistance of the bulk sample changes, the material must either have the conductivity distributed unevenly or the change of resistance with strain must vary. It has been shown before [4] that the latter is not the case for deformations in elastic regime.

To corroborate the first case the two layer beams were produced. The results for these samples show the effect more clearly, and are consistent with the previously mentioned theories. Slight deviations from the expected behavior such as the observable hysteresis can be attributed to local stress concentrations e.g. at the supports of the setup.

Additionally a detailed characterization of the sample material was conducted. This is done to support the thesis, that the production process directly influences this behavior. A layer by layer analysis of the material showed that the conductivity distribution is indeed dependent on the production process. The raster sampling showed a pronounced influence of the proximity of boundaries that can reduce material flow or act as heat sources or sinks. These distributions should also be visible in additional experiments designed to show in-plane variations of conductivity. It has become clear that the curing process can also influence the conductive network formed in a bulk sample significantly, e.g. through convection or variations in viscosity [6]. A characterization of the conductivity variations at different scales as well as microscopic characterization of the material is planned.

Furthermore, the local conductivity of the particle network is dependent only on the concentration but also its microstructure. These parameters are to be determined in future work as well.

CONCLUSIONS

In this work a simple process for the production of directional sensitive nanocomposite strain sensors was presented. The sensors show an approximately linear and reversible change of electrical resistance when subjected to elastic deformations. It was shown that a conductivity gradient/distribution can easily be induced through various parameters in the production process. Furthermore approaches for a closer control of the resulting properties have been presented.

ACKNOWLEDGMENTS

The German Federal Department for Research and Education is gratefully acknowledged for financial support of the project 03X0042M "Verbundprojekt: Dispergierung und Konfektionierung" (CarboDis).

REFERENCES

1. C. Li et al. / *Composites Science and Technology* 68 (2008) 1227–1249

2. Sandler J K W, Kirk J E, Kinloch I A, Shaffer M S P and Windle A H (2003) *Polymer* **44** 5893–9

3. Böger L, Wichmann M H G, Meyer L O and Schulte K (2008) *Compos. Sci. Technol.* **68** 1886–94

4. M H G Wichmann, S T Buschhorn, L Böger, R Adelung and K Schulte (2008) *Nanotechnology* **19** 475503

5. Filonenko-Borodich, M. 1964. *Theory of elasticity*. Groningen: P. Noordhoff

6. Battisti A et al., Monitoring dispersion of carbon nanotubes in a thermosetting polyester resin, *Compos Sci Technol* (2008), doi:10.1016

Mater. Res. Soc. Symp. Proc. Vol. 1134 © 2009 Materials Research Society 1134-BB08-04

Fabrication of an Electrochromic Device Based on Polyaniline-Poly(vinyl Alcohol)-Natural Polymer Blends

Michael I. Ibrahim, Maria J. Bassil, Mario R. El Tahchi, Joseph K. Farah

GBMI, Department of Physics, Lebanese University – Faculty of Sciences 2, POBox 90656 Jdeidet, Lebanon, e-mail: gbmi@ul.edu.lb, Tel: +961 3 209688, Fax: +961 1 681553

ABSTRACT

Polyaniline (PANI), discovered over 150 years ago, is a good electrochromic (EC) material with high electrical conductivity (up to 5 S/cm under optimized conditions). It is simple to polymerize since it can be prepared chemically or electrochemically by the oxidation of aniline and has a good chemical stability. Based on PANI, the fabrication of an EC flexible layer, which can be used as an EC device in buildings and cars windows, is possible. To overcome the poor mechanical properties of PANI, it can be mixed with polymer and non-polymer materials. Relying on the properties of PANI, the present study consists of preparing blends of three polymers which are poly(vinyl alcohol) PVA, PANI and a natural polymer to produce an electrically conducting layers which can change their color by applying a low voltage, at the same time these layers will have the property of light polarizer. Thus the control of the transmission of these layers will be either by using the polarized light activity or based on the EC effect.

INTRODUCTION

Electrochromism is a property of a material, organic or inorganic, consisting in changes of the absorption spectrum occurring upon application of an electrical stimulus. In most cases the mechanism of this effect is electrochemical or oxidation/reduction reaction. Many inorganic materials have EC properties such as tungsten oxide WO_3 [1-3], Prussian blue [4, 5], and other inorganic materials [6]. Color changes are commonly between a transparent or bleached state or between different colored states.

Due to the electrochemical reaction of PANI, it becomes fully reduced at the positive metal electrode and switches to the leucoemeraldine state (faint yellow). PANI becomes half oxidized at the negative metal electrode and switches to the emeraldine base state (blue) followed by the pernigraniline state (dark violet color or black and presenting low conductivity) [7, 8].

The pathways of the different states of PANI can be resumed in the following oxidation/reduction reaction [9]:

$$(C_{24}H_{18}N_4\text{-}2H)_n \xleftarrow{+2e^- -2Cl^-} (C_{24}H_{18}N_4\text{-}2HCl)_n \xrightarrow{-2(H^+ +Cl^-)} (C_{24}H_{18}N_4)_n \xrightarrow{-2(H^+ +e^-)} (C_{24}H_{16}N_4)_n$$

leucoemeraldine emeraldine salt emeraldine base pernigraniline

To overcome the poor mechanical properties of PANI, PANI can be mixed with polymer and non-polymer materials to form blends and composites. In blends, the conducting PANI particles will be dispersed in an insulating polymer matrix. Many papers have studied the conductivity of PANI blends. For example PANI-PVA with methyl cellulose MC blends present semiconducting behavior independently of MC volume fraction. In these blends the conductivity decreases with the decreasing volume fraction due to the superlocalization of the electronic states in the structure [10, 11]. Other blends could be made with PANI and polymers with different methods of polymerization [7].

In this work, PANI powder, PANI-PVA and PANI-cellulose are studied. The charge transfer resistance is measured and charge transfer conductivity is determined. The speed of transition from one state to another is presented for three types of large scale samples. The electrical current is monitored to determine the physical process taking place inside these samples.

EXPERIMENT

PANI can be prepared using different methods such as electrochemical polymerization of electrolyte consisting of aniline and hydrochloric acid (HCl) using cyclic potential techniques, or the chemical oxidative polymerization. The chemical oxidative polymerization consists on polymerizing an aqueous solution of aniline-HCl (C_6H_7N-HCl) with ammonium peroxydisulfate ((NH_4)$_2S_2O_8$). In this study, PANI was first prepared using different concentrations of aniline-HCl and ammonium peroxydisulfate.

The standard preparation of PANI [12] was used to prepare the PANI powder samples and a PANI standard solution. Aniline hydrochloride, 2.59g , (Sigma-Aldrich purum $\geq$99%) was dissolved in distilled water to 50 ml of solution and 5.71g ammonium peroxydisulfate (Fluka purum $\geq$98%) was dissolved is distilled water to 50 ml of solution. The polymerization starts when the two solutions were mixed together, and the temperature of the mixture increases. After ten minutes green emeraldine salt PANI is formed. The chemical reaction of PANI's formation is:

$$4n\ (C_6H_7N\text{-HCl}) + 5n\ (NH_4)_2S_2O_8 \longrightarrow (C_{24}H_{18}N_4\text{-2HCl})_n + 5n\ (NH_4)_2SO_4 + 2n\ HCl + 5n\ H_2SO_4$$

where n is the degree of polymerization.

From the obtained PANI aqueous solution (PANIaq), 10g were dried in the oven at a temperature of 150°C from 1.5 to 2 hours in order to obtain 0.694g of PANI powder. To test the electrochromic activity of PANI powder, PANI standard solution was placed on a glass substrate and dried at room temperature for 2 to 3 hours (sample P0). To prepare the PANI-PVA blends three methods were used. The first consists on mixing PANI powder (2 to 5% of the total solution weight) with a 10% aqueous solution of PVA (BP-24, 86% hydrolyzed, Chang Chun Petrochemical Co. Ltd). The polymerization of aniline taking place inside a PVA aqueous solution defines the second method. Under these conditions, five samples labeled from I-a to V-a were prepared using 1ml of 10% PVA, 2.59% of aniline-HCl and 5.71% of ammonium peroxydisulfate and different concentrations of distilled water (0, 48.1, 65, 73.4, and 79% respectively). The third method consists on mixing PANI standard solution with PVA 10%. Five samples labeled from I-b to V-b were prepared using a mixture of 1ml of 10% PVA, and different concentrations of PANIaq (50, 66, 75, 80, and 83.3% respectively). PANI-cellulose samples were prepared by using cellulose sheets of thickness 25 μm. PANI solution was prepared using the standard technique but DMF was used instead of water. PANI in DMF was then spread on the cellulose sheet clamped on a glass substrate and left to dry for 12 hours (sample C0). The excess of PANI, formed on the surface of the samples is then wiped off. The sample humidity was then increased, while clamped, by placing the sample in a steam bath for 5 minutes.

The current-voltage characteristics of the samples are measured using a Keithley 487 pico-ammeter/voltage source. The sample is clamped between two aluminum electrodes and placed in an aluminum protective box to reduce external electromagnetic interferences (figure 1-a). To follow the electrochromic activity and the ion transfer in PANI samples, the setup in figure 1-b was used. An ammeter recorded the current while a CCD camera recorded the color variations of PANI samples.

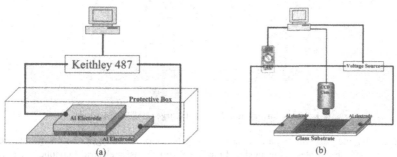

(a)   (b)

Figure 1: Sample holder used to measure electrical conductivity of PANI (a) and electrochromic activity and the ion transfer (b).

DISCUSSION

PANI powder sample mixed with PVA showed aggregations of PANI when dried (Figure 2), for that reason these samples where not characterized, either electrically (conductivity) or optically (EC activity).

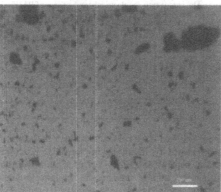

Figure 2: Aggregation of PANI powder in PVA.

The current density, j, versus the applied voltage, V, for samples I-a to V-a are shown on figure 3. The characteristics of samples I-b to V-b and of samples P0 and C0 are of similar shape, and are not shown here, for clarity.

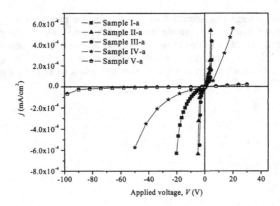

Figure 3: Current density versus the applied voltage for all PANI I samples.

Fitting linearly the most linear part of the curves at the positive voltages, the charge transfer resistance R_{ct} of the samples is obtained and thus the charge transfer electrical conductivity, σ_{ct}, given by the following relation:

$$\sigma_{ct} = \frac{t}{R_{ct} l w} \tag{1}$$

where t is the thickness, l the length and w the width of the sample. The electrical conductivities of the samples are shown in table 1.

Table 1: The charge transfer electrical conductivity of PANI samples.

Sample Number	σ_{ct} (S/cm)	Sample Number	σ_{ct} (S/cm)
I-a	0.3631×10^{-6}	I-b	2.859×10^{-7}
II-a	4.166×10^{-6}	II-b	3.218×10^{-7}
III-a	2.071×10^{-6}	III-b	3.66×10^{-7}
IV-a	0.1625×10^{-6}	IV-b	3.866×10^{-7}
V-a	0.288×10^{-8}	V-b	8.625×10^{-7}
P0	6.949×10^{-6}	C0	2.707×10^{-7}

Table 1 shows that all the PANI samples have conductivities in the semiconductor range. Sample II-a shows the highest conductivity in all I-type samples. From sample III-a to V-a, the conductivity decreases (2 to nearly 1450 times that of sample II-a) due to the increase of the water concentration during the sample preparation, leading to higher dispersion of PANI molecules inside the mixture. The behavior of these samples tends toward an insulator (PVA) behavior. Sample I-a conductivity is 11 times lower than that of sample 2 due to the lack of the appropriate quantity of solvent and the polymerization is not completely finished.

For PANI samples b, they show nearly the same conductivity values, which are lower that the "a" samples. This is due to the particle size of PANI, since it was polymerized separately in water. Note that initial mixtures of all samples were thoroughly shaken for 5 minutes, in order to disperse all the PANI particles inside the PVA matrix.

The electrochromic activity is characterized by measuring the speed of diffusion of the leucoemeraldine state (Figure 4-b) and by monitoring the variation of electric current density, j, with respect to the elapsed time, t (figure 4-a). For all the samples, a potential difference of 35 V is applied between two aluminum electrodes, connected to the edges of the sample as shown in figure 1-b. Sample P0, having highest conductivity, induces the highest start-up current and the fastest diffusion speed (average of 6.5 mm/min). PANI-PVA (V-b) shows lower EC activity and an average speed of 1.2 mm/min. The PANI-cellulose samples showed lower conductivity (table 1) and the ion diffusion appears to be the slowest in all the studied samples (figure 4-a). This is due to the polymerization of PANI inside the cellulose, after the aniline-HCl in DMF was spread on the cellulose sheet. The humidity stays the main parameter to control in these samples, thus the electrochromic activity could not be monitored carefully.

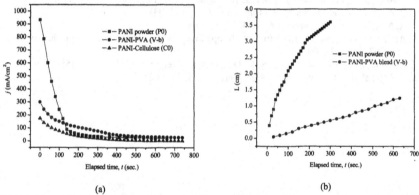

(a)   (b)

Figure 4: The variation of electric current density, j, with respect to the elapsed time, t (a) and the speed of diffusion of the leucoemeraldine state (b).

CONCLUSIONS

Polyaniline electrical properties and electrochromic activity were studied. Several methods of preparation of electrochromic layers were presented. The volume ratio of PANI to PVA is showed to be an important parameter. In addition, the solvent quantity affects the reaction end-results and the dispersion of PANI in PVA. The samples prepared by mixing the standard solution with PVA with different volume ratio show lower conductivity than those prepared by polymerizing Aniline-HCl in PVA. The architecture of low cost EC devices is possible using PANI-PVA blend that acts like solid electrolyte. The use of PANI-cellulose is reported here as a preliminary study, further research are needed to define the optimum operating conditions of these samples.

ACKNOWLEDGMENTS

This work was carried out thanks to the financial support of the Lebanese National Council for Scientific Research and the Lebanese University.

REFERENCES

1. C. Granqvist, *Jour. Euro. Cer. Soci.* 25, 2907 (2005).
2. F. Jiang, T. Zheng and Y. Yang, *Jour. Non-Cryst. Sol.* 354, 1290 (2008).
3. Y. Lu and C. Hu, *Jour. All. Comp.* 449, 389 (2008).
4. B. Jelle, G. Hagen, and S. Nodland, *Electrochimica Acta* 38, 1497 (1993).
5. M. Carpenter and R. Conell, *Appl. Phys. Let.* 55, 2245 (1989).
6. A. Argun, PhD. Thesis, University of Florida, 2004.
7. J. Anand, S. Palaniappan, and D. Sathyanarayana, *Prog. Poly. Sci.* 23, 993 (1998).
8. P. Andersson and M. Berggren, *Appl. Phys. Let.* 83 (2003).
9. R. Mortimer, *Electrochimica Acta* 44, 2971 (1999).
10. A. Sarkar, P. Ghosh, A. Meikap, S. Chattopadhyay, S. Chatterjee and M. Ghosh, *Jour. Appl. Phys.* 97, 113713 (2005).
11. P. Dutta, S. Biswas, M. Ghosh, S. K. De and S. Chatterjee, *Synt. Met.* 122, 455 (2001).
12. J. Stejskal and R. Gilbert, *Pure Appl. Chem.* 74, 857 (2002).

Mater. Res. Soc. Symp. Proc. Vol. 1134 © 2009 Materials Research Society   1134-BB08-10

Dielectric Response of Ceramic-Polymer Composite With High Permittivity

Xiaobing Shan, Lin Zhang, Peixuan Wu, Canran Xu and Z.-Y Cheng
Materials Research and Education Center, Auburn University, Auburn, AL 36849

ABSTRACT

Based on solution casting method, a ceramic [$CaCu_3Ti_4O_{12}$ (CCTO)]-Polymer [P(VDF-TrFE)] composite with flexibility has been synthesized and its dielectric response has been studied. The CCTO ceramic powders were prepared by traditional sintering method and were milled with a relative uniform size. The dielectric properties of these films with micro-size and nano-size CCTO particle, as well as different polymer matrixes were determined. The process was optimized by hot pressing and surface modification to achieve high dielectric constant. A dielectric constant about 175 for one layer composite with high flexibility using silane coupling agent was obtained at 1 kHz at room temperature.

INTRODUCTION

In general, inorganic ceramics have high dielectric constants of up to 10^4. However, they are brittle and heavy, and have low electric breakdown strengths. In contrast, polymers are flexible and easy to process and exhibit high breakdown strengths (up to 500 MV/m), but their dielectric constants are much smaller (<10). Therefore, in the last two decades, a great amount of efforts has been given to the development of so-called 0-3 composites to fully take the advantages of ceramic and polymers [1-3]. In those 0-3 composites, ceramic powders are used as filler and polymers are used as matrix. A common method to acquire high dielectric constant in the composites is to utilize the powders of ceramics with high dielectric constant, which are mostly ferroelectric-related ceramics, such as $Pb(Mg_{1/3}Nb_{2/3})O_3$, and $Pb(Zr,Ti)O_3$ [4-5]. However, the ferroelectric ceramics exhibit a high electromechanical coupling effect, which may result in a considerable mechanical strain when the composite is under strong electric field. A high electric-field-induced mechanical strain for many dielectric applications is not desirable.

In this paper, 0-3 composites using powders of high-dielectric-constant ceramics $CaCu_3Ti_4O_{12}$ (CCTO) are developed. The CCTO exhibits a very high dielectric constant and does not have the piezoelectric effect. Therefore, the CCTO is of very weak electromechanical coupling effect. As presented below, the CCTO-based composites do exhibit a high dielectric constant that is higher than the dielectric constant observed in other 0-3 ceramic-polymer composites. The dielectric properties with micro-size and nanosize CCTO particle, as well as different polymer matrixes were studied.

EXPERIMENTAL

The CCTO ceramics were prepared using the conventional powder processing method. High purity metal oxide powders of Calcium Carbonate ($CaCO_3$, 99.5 %, Alfa Aesar), Copper

Oxide (CuO, 99.7 %, Alfa Aesar) and Titanium Dioxide (TiO$_2$, 99.8 %, Alfa Aesar) were used to prepare the ceramics. The mixture of these chemicals was milled and then calcined at 900~1075 °C in Al$_2$O$_3$ crucible. After calcinations, the ceramic powers were sintered at 1075 °C for different times. In the research, P(VDF-TrFE) 55/45 vol.% copolymer was utilized as the polymer matrix for the fabrication of the 0-3 composites. As P(VDF-TrFE) copolymers, P(VDF-TrFE) 55/45 vol.% exhibits a relative higher dielectric constant than other polymers. At the same time, P(VDF-TrFE) 55/45 vol.% copolymer is of a very weak electromechanical coupling effect comparing with other P(VDF-TrFE) copolymers.

The CCTO-P(VDF-TrFE) 0-3 composite were prepared by traditional solution casting method followed by a hot-pressing techniques. First of all, P(VDF-TrFE) 55/45 mol% copolymer was dissolved in dimethyl formamide (DMF) under magnetic stirring for 5 hours. Then, the CCTO powders were added into the solution that was then stirred for 8 hours and followed by 20 minutes sonication. The final CCTO- P(VDF-TrFE)-DMF solution/suspension was casted on a glass plat at 70 °C for 8 hours. However, it is found that the uniformity of the as-cast composites is not good as described in our previous papers [6]. That is, in the as-cast composite film, one side is polymer-rich with a almost pure polymer layer.

To improve the uniformity of the composites, the as-cast composite film was then hot pressed at 200 °C. During the process, multiple layers of the as-cast composite film were stacked together using two different configurations. In one configuration that is named as PP, the polymer-rich side of an as-cast composite film was placed to face the polymer-rich side of another as-cast film. In the other configuration that is named as PC, the polymer-rich side of an as-cast composite film faced the non-polymer-rich side of another as-cast composite film. In order to improve the crystallinity of the polymer matrix, an annealing process at 125 °C was carried out for the hot-press composites.

To characterize the dielectric properties of the composites, gold film was sputtered on the both sides of the sample to serve as the electrodes. An Agilent 4294A impedance analyzer was employed to measure the impedance spectrum of the sample at frequencies from 100 Hz to 1 MHz. The dielectric properties of the samples were calculated from the impedance using parallel capacitor model. JEOL (JSM-7000F) Field Emission Scanning Microscope was used to examine the morphology and the microstructure of the composites.

RESULTS AND DISCUSSION

Effect of hot pressing processing

The dielectric constants of the 0-3 composites with 40 vol% CCTO prepared using different conditions, including hot-pressing (HP) or not and annealing or not, are shown in Figure 1. During the hot-pressing process, the as-cast composite film was stacked using PC configuration. Clearly, the dielectric constant of the final composites is strongly dependent on the number of the as-cast composite layers used in the hot pressing process and their annealing conditions. However, it seems that there is a saturated value in the dielectric constant.

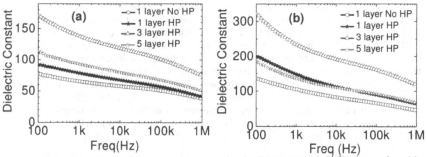

Figure 1. The dielectric constant versus frequency for the CCTO-P(VDF-TrFE) composite with 40 vol.% CCTO at room temperature. The composites with different numbers (1, 3, and 5) of as-cast composite layers during hot pressing (HP) were presented. During the HP process, the PC configuration was used. In the figure the composites without HP (1 layer No HP) are also included. (a) composites without annealing at 125 °C, and (b) composites annealing at 125 °C for 8 hours.

For the composites by using PP configuration during HP process, the dielectric properties were characterized. It is found that the dielectric constant of those composites is much higher than the composites made using PC configuration during HP process and that the HP process, especially the time, is very critical as shown in Table 1.

Table 1. Summery of dielectric constant (κ) and dielectric loss (tanδ) at 1 kHz of CCTO-P(VDF-TrFE) composites at room temperature. Here the composites are of 50 vol.% CCTO and were prepared using the PC configuration during the HP process. During HP process, different times (10, 20, and 30 seconds, respectively) were used.

	Hot compression (HP) time					
	10 s		20 s		30 s	
	κ	tgδ	κ	tgδ	κ	tgδ
2 Layer	167	0.2	147	0.3	163	0.3
4 Layer	95	0.1	162	0.2	234	0.3
6 Layer	205	0.1	332	0.2	474	0.2

Effect of ceramic size and polymer matrix

The dielectric constants of the 0-3 CCTO-P(VDF-TrFE) composites with 50 vol% nano-size CCTO, including hot-pressing (HP), are shown in Figure 2. As increasing the hot pressing layers, the dielectric constant has been improved from 37 to 53 in Figure 2 (a). In order to optimize the dielectric results, composite sample has been prepared with different hot pressing time, such as 10, 20, 30, and 40s. Based on the experimental results in Figure 2 (b), it is found that there is a saturated value that the dielectric constant increases from 53 to 59, then it reached a critical point that the value falls down to 38. All the experimental results show the hot pressing has played an important role in optimizing the dielectric response of the composite. At the same time, the effect of different polymer matrix been studied. The dielectric constants of 0-3 CCTO-P(VDF-CTFE) composite are shown in Figure 2 (c) and it can be observed that the dielectric property can be optimized and a relative high dielectric constant about 82 has been reached, which is about 54% higher than the maximum value in Figure 2 (1).

It should be noted that after comparing with the results in table 1, all the results using nano-size CCTO in different polymer matrix are relative low and it can be explained with its SEM image in Figure 2 (d). After studying the morphology of the composite in Figure 3, there is a ceramic rich segregation region around 10 ηm in width in the middle which lead to low dense structure and poor dielectric performance.

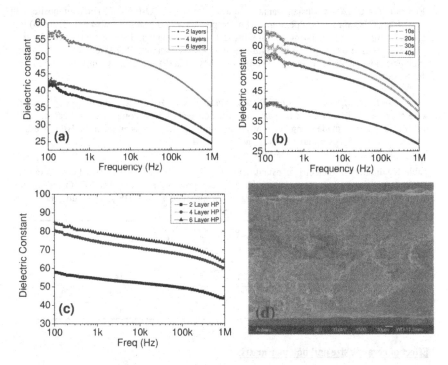

Figure 2. The dielectric constant versus frequency for the composite with 50 vol.% CCTO at room temperature. (a) CCTO-P(VDF-TrFE) composites HP for 2, 4, 6 layers, and (b) CCTO-P(VDF-TrFE) composites HP for 10, 20, 30, 40s. (c) CCTO-P(VDF-CTFE) composite HP for 2, 4, 6 layer. (d) SEM image of CCTO-P(VDF-TrFE) composite with 50 vol% nano-size CCTO

Effect of silane coupling agent

The dielectric properties of the 0-3 CCTO-P(VDF-TrFE) composites with 0.3, 0.5, 0.75, 1, 5 and 10 wt% silane have been studied in this work. The dielectric results with 0.5 wt% silane are shown in Figure 3. Base on the results in Figure 3 (a), one layer sample with high flexibility and high dielectric constant as high as 175 has been fabricated and the dielectric results for multiple layers are listed in table 2. The bridge-linked action of silane molecule would effectively enable the transfer of electrical polarization in the interfacial layer between CCTO and its polymer matrix and it could result in the unexpected dielectric properties. All the results suggest they possess potential applications as a new dielectric material.

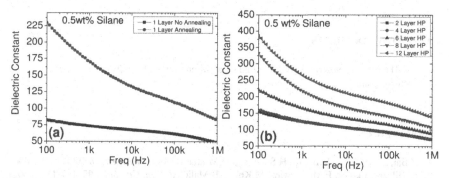

Figure 3. The dielectric constant versus frequency for the composite with 50 vol% CCTO at room temperature. (a) Annealing effect for CCTO-P(VDF-TrFE) composites , and (b) CCTO-P(VDF-TrFE) composites HP for 2, 4, 6, 8, 12 layers.

Table 2. Summery of dielectric constant (κ) and dielectric loss ($\tan\delta$) at 1 kHz of CCTO-P(VDF-TrFE) composites with 0.3, 0.5, 0.75, 1, 5 and 10 wt% silane at room temperature

	Silane coupling agent concentration					
	0.3wt%	0.5wt%	0.75wt%	1wt%	5wt%	10wt%
1 layer	64/0.07	170/0.20	114/0.21	67/0.156	159/0.21	85/0.308
2 layers	120/0.14	125/0.133	115/0.36	65/0.125	305/1.44	99/0.208
4 layers	137/0.29	126/0.147	146/0.29	78/0.158	444/1.22	128/0.38
6 layers	145/0.15	166/0.18	281/0.40	87/0.147	394/0.979	191/0.449
8 layers	141/0.16	218/0.268	231/0.34	81/0.158	394/0.97	143/0.36
12 layers	128/0.16	267/0.228	208/0.35	80/0.159	318/0.94	106/0.225

CONCLUSIONS

Ceramic-polymer 0-3 composites using CCTO powders as filler are introduced. It is found that those composites exhibit a very high dielectric constant. More interestingly, it is found that the dielectric response of the composites is strongly dependent on the composite preparation process, such as whether the hot-pressing process was employed, the number of the as-cast composite film used in the hot-pressing process, the configuration of the as-cast composites film in the hot-pressing process, the time used in the hot-pressing process, as well as whether the composites were thermally annealed. All those indicate that the high dielectric constant observed in the CCTO-P(VDF-TrFE) composites is an intrinsic effect. That is, the interfacial layers between the CCTO powders and polymer matrix are critical.

ACKNOWLEDGMENTS

The research was partially supported by a 3M non-tenured faculty grant and an AU competitive grant.

REFERENCES

1. Y.Bai, Z.-Y Cheng, V.Bharti, H.S.Xu, and Q.M.Zhang: *Appl.Phys.Lett.* **76**, 3804-380 (2000)
2. B. Hilczer, J.kulek, E.Markiewicz, M.Kosec, B.Malic: *J. Non. Cry. Sol.* **305**, 167-173 (2002)
3. Baojin Chu, Xin Zhou, Kailiang Ren, Bret Neese, Minren Lin, Qing Wang, F.Bauer, Q.M.Zhang. *Science*, **313**, 334-336 (2006)
4. Feng Xia, Z.-Y. Cheng, H.S. Xu, H.F. Li, Q.M. Zhang, G.J. Kavarnos, R.Y. Ting, G. Abdul-Sedat, and K.D. Belfield, *Adv. Mater.* **14**, 21, 1574-1577 (2002).
5. Z.-Y.Cheng, V.Bharti, T.-B. Xu, Haisheng Xu, T.Mai, Q.M.Zhang. *Sensors and actuators* A **90**, 138-147 (2001)
6. M. Arbatti, X.B Shan, and Z.-Y.Cheng: *Adv. Mater.* **19**, 1369-1372 (2007).

Mater. Res. Soc. Symp. Proc. Vol. 1134 © 2009 Materials Research Society   1134-BB08-11

Crystalinity Properties of Carbon Nanotube-Polyvinylidene Fluoride Composites

Xiaobing Shan, Peixuan Wu, Lin Zhang and Z.-Y Cheng
Materials Research and Education Center, Auburn University, Auburn, AL 36849

ABSTRACT

Single-wall and multi-wall carbon nanotube blends (0 to 0.5 vol%) with polyvinylidene fluoride (PVDF) have been prepared using solution cast method and characterized. By acid treatment, it has been observed that nanotube has been well functionalized and uniformly dispersed into the polymer. X-ray diffraction analysis coupled with differential scanning calorimetry (DSC) has revealed that carbon nanotube alters the crystallinity of PVDF and thereby enhances the β-phase in PVDF. Experimental results have demonstrated that enhancement of β-phase is a function of carbon nanotube concentration.

INTRODUCTION

It is well known that due to its novel structural, electronic, and mechanical properties, carbon nanotube has been applied in many fields, such as carbon nanotube composites. Incorporation of carbon nanotube into the organic polymers enables the fabrication of new high performance hybrid materials with good electrical and thermal transport properties [1-3]. A significant increase in thermal conductivity of organic fluid and polymers with relative low concentration of carbon nanotubes has been reported [4-7]. Poly(vinylidene fluoride) (PVDF) has attracted much attention because of its widely applications to high pyroelectric, piezoelectric, and ferroelectric materials. It is known that PVDF exhibit four crystalline phases, namely α, β, γ and δ phases [8]. In those phases, β-phase is especially interest due to its large pyro and piezoelectric effect. Currently, the mainly interest is focused on β-phase PVDF [9-10]. Comparing with P(VDF-TrFE), PVDF homopolymer contains more dipoles in its molecular structure and is at much lower cost. However, α-phase is thermodynamically more stable in PVDF than β-phase. The influence of CNTs on α –β phase transition in PVDF has been investigated by adjusting the CNTs concentration and processing parameters.

In this paper, carbon nanotube- Poly(vinylidene fluoride) composites are fabricated and we investigate functionalized carbon nanotubes/ Poly(vinylidene fluoride) (SWNT or MWNT/PVDF) composites with low concentrations of SWNT or MWNT (0 to 0.5 vol%). X-ray diffraction analysis and DSC has been used to study its properties, such as crystallinity, at different annealing temperatures.

EXPERIMENTAL

The carbon nanotube (SWNT or MWNT) were obtained from a commercial source (Shenzhen Nanotech Port Co. Ltd., China). A commercial SOLEF 1010 Poly(vinylidene fluoride) was supplied from Solvay Solexis, Inc. The carbon nanotube-polymer composite were prepared by traditional casting method and hot pressing techniques. The carbon nanotube was functionalized before mixed with PVDF. The functionalized procedure is shown as following:

1. 3:1 Concentrated H_2SO_4: HNO_3 mixture was chosen for the oxidizing acid in the cutting operation.
2. After weighting the CNT, Stirring the mixture of CNT and concentrated acid for amount of time by stirring bar and sonicator (Heat system-ultrasonics INC)
3. Using SiO_2 filter to separate CNT
4. Washing by Hydrochloric Acid
5. Washing by distilled water until the PH=7
6. Dry at 70 ^oC/13Hr in vacuum (76 cm Hg)

Those nanotube composite were prepared using traditional casting method by dissolving poly(vinylidene fluoride) (PVDF). The solution was stirred for 8 hours and then sonicated for about 20 mins in order to disperse CNTs uniformly inside copolymer.

Poly(vinylidene fluoride) (PVDF) was dissolved in dimethyl formamide (DMF) under magnetic stirring for 5 hours. Functionalized carbon nanotube (CNTs) with 0.01, 0.02, 0.03, 0.05, 0.1, 0.5 vol % SWNT or MWNT was then dispersed into the solution and then it was stirred continuously for 48 hours and sonicated for about 20 mins in order to distribute CNT uniformly. The final carbon nanotube- PVDF solution was casted on a glass plat at 40 °C for 8 hours. In order to improve the crystallinity, further annealing processing has been applied.

Gold electrode was sputtered on both sides of the samples. Agilent 4294A impedance analyzer was employed to characterize the dielectric property of the samples. The measured frequency range is 100 Hz to 1000 kHz. JEOL (JSM-7000F) field emission scanning microscope was used to examine the surface morphology. The thermal properties and phase transition was conducted by using TA DSC system (model 2910). In DSC measurement, 10 mg sample was weighted, and then was heated at 10 °C/min to 200 °C, followed by cooling down at 10°C/min to room temperature. The microstructure of the crystalline phase in the blend was characterized by X-ray (Rigaku system) to determine the (200, 110) peak. Thermal gravimetry analysis (TGA 2050, TA instruments) was employed to quantify the wt% of organic part in functionalized-CNT.

RESULTS AND DISCUSSION

CNT Functionalization

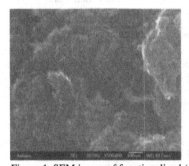 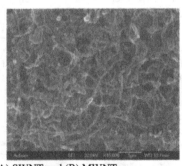

Figure 1. SEM image of functionalized (A) SWNT and (B) MWNT

All of the current known forms of CNTs are insoluble in organic solvents, making it hard for further application. In the experiment, all the CNTs have been functionalized by acid treatment in order to attach –COOH group to CNTs as shown in Figure 1 and the corresponding TGA results are shown in Figure 2. It is known that the observed loss of the unmodified CNTs from TGA is believed to be small amounts of amorphous carbon and impurities. If discounting the mass loss of carbon nanotubes, the mass fraction of organic groups lost at 600 °C was 5 wt% for SWNT and 19 wt% for MWNT. These values indicated that the surface modification process led to a significant mass fraction of organic groups chemically bonded to CNTs.

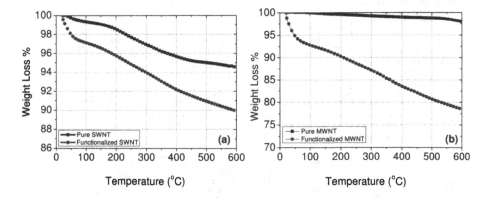

Figure 2. TGA thermograms of neat and functionalized SWNT or MWNT.

CNT-Differential Scanning Calorimetry (DSC)

The thermal properties in CNTs-PVDF composites with 0.01, 0.02, 0.03, 0.05, 0.1, 0.5 vol % SWNT or MWNT have been studied by the differential scanning calorimetry (DSC) at different annealing temperature. The DSC results are shown in Table 1. The fraction of enthalpy improved by incorporation of MWNT is 17% for annealing at 125 °C, 21% for annealing at 135 °C, and 25 % at 140 °C. Similarly, it is interesting to find that the same phenomena were observed in other annealing temperature, such as 125 °C and 135 °C. Corroborated with results of increasing melting point, all the experimental results demonstrated that CNTs promote the heterogeneous nucleation within the PVDF matrix and can be used as a nucleating agent to improve the crytalinity of PVDF.

Table 1. Summery of DSC results for MWNTs-PVDF composite

	MWNTs vol %	Peak (°C)	Enthalpy (J/g)
Pure *PVDF* casting at 40 °C	0	160.94	60.5
Pure *PVDF* casting at 40 °C and annealing at 125 °C	0	160.00	63.86
MWNTs-PVDF casting at 40 °C and annealing at 125 °C	0.01	162.84	63.84
	0.02	161.28	63.49
	0.03	161.01	63.62
	0.05	159.96	73.45
	0.1	159.70	73.14
	0.5	159.61	71.11
Pure *PVDF* casting at 40 °C and annealing at 135 °C	0	159.14	57.57
MWNTs-PVDF casting at 40 °C and annealing at 135 °C	0.01	161.96	63.14
	0.02	161.27	69.31
	0.03	160.89	64.09
	0.05	160.18	61.57
	0.1	159.34	61.54
	0.5	159.42	61.55
Pure *PVDF* casting at 40 °C and annealing at 140 °C	0	160.38	67.35
MWNTs-PVDF casting at 40 °C and annealing at 140 °C	0.01	160.86	58.02
	0.02	160.43	70.56
	0.03	160.00	84.30
	0.05	160.22	66.77
	0.1	159.52	67.16
	0.5	159.29	70.26

X-ray diffraction

· X-ray diffraction was used to observe the effect of CNTs contents on the microstructure of PVDF. Based on XRD experimental data in Figure 3, the diffraction peak at 18~19° characteristic of the α-phase is tending to the 20~21° peak, characteristic of the β-phase. The results have exhibited decreased peaks for α-phase as those SWNT or MWNT-PVDF composites has been treated at 40 °C and 155 °C. Especially in comparison with Figure 3 (a) and (b), it can be found that MWNT-PVDF composite results in more apparent α-phase to β-phase transition. In order to characterize this transition in detail, all the XRD results have been analyzed with professional peak-fitting software and the fitting results of XRD peaks are listed in Table 2. Based on experimental results, it should be noted that the β phase increasing from 61% to 71% as casting at 40 °C and 69% to 71% as annealed at 155 °C, and the most enhanced results are achieved with 0.01 to 0.03 vol% MWNT within the PVDF polymer matrix.

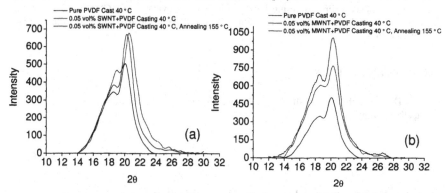

Figure 3. X-ray diffraction pattern of (a) SWNT-polymer composite; (b) MWNT-polymer composite.

Table 2. Summery of XRD fitting results

	CNTs vol %	Peak 2θ	α Area (Percentage)	β Area (Percentage)
PVDF Casting at 40 °C	0	17.6/19.8	39	61
MWNT-PVDF casting at 40 °C	0.01	18.2/20.7	29	71
	0.02	17.7/20.1	29	71
	0.03	18.4/20.8	26	74
	0.05	17.5/20.1	33	66
	0.1	17.3/20.0	32	67
	0.5	17.4/20.1	40	60
PVDF casting at 40 °C and then annealing at 155 °C	0	17.7/20.2	31	69
MWNT-PVDF casting at 40 °C and then annealing at 155 °C	0.01	17.8/20.3	27	73
	0.02	17.8/20.3	29	71
	0.03	17.7/20.3	29	71
	0.05	17.6/20.3	35	65
	0.1	17.3/20.3	34	66
	0.5	17.6/20.3	35	65

CONCLUSIONS

The CNTs-PVDF composites have been fabricated by the traditional casting method. The influence of CNT on $\alpha - \beta$ phase transition in PVDF has been investigated by adjusting the CNT concentration (0.01, 0.02, 0.03, 0.05, 0.1, and 0.5 vol% CNT) and processing parameters. In order to disperse uniformly into the polymer matrix, CNTs have been functionalized first by acid-treatment and TGA results have confirmed that better distribution of CNTs has been achieved. All the experimental results have provided strong evidence that CNTs can promote the heterogeneous nucleation within the PVDF matrix and CNTs can lead to apparent α-phase to β-phase transition.

ACKNOWLEDGMENTS

The research was partially supported by a 3M non-tenured faculty grant and an AU competitive grant.

REFERENCES

1. Y.Zou, Y.Feng, L.Wang, X.Liu, *Carbon.* **42**, 271-277 (2004)
2. B.G.Min, T.V.Sreekumar, T.Uchida, S.Kumar, *Carbon.* **43**, 599-604 (2005)
3. T. Furukawa, *Key Eng. Mater.* **15**, 92-93 (1994)
4. I.Alexandron, E.Kymakis, and G.A.J.Amaratunga, *Appl.Phys.Lett.* **80**, 1435-1437 (2002)
5. M. J. Biercuk, M. C. Llaguno, M. Radosavljevic, J. K. Hyun, A. T. Johnsond, and J. E. Fischer, *Appl. Phys. Lett.* **80**, 2767-2769 (2002).
6. S. U. S. Choi, Z. G. Zhang, W. Yu, E. A. Lockwood, and E. A. Grulke, *Appl. Phys. Lett.* **79**, 2252-2254 (2001).
7. Ce-Wen Nan, Gang Liu, Yuanhua Lin, and Ming Li, *Appl. Phys. Lett.* **85**, 3549-3551 (2001).
8. K.C. Kao in *Dielectric Phenomena in Solid*, (San Diego, Elsevier Academic Press, 2004) pp. 234-236.
9. X.He, K.Yao, *Appl. Phys. Lett.* **89**, 112909-3 (2006).
10. C.J.L.Constantino, A.E.Job, R.D.Simocs, J.A.Giacometti, V.Zucolotto, O.N.Oliverira, JR., G.Gozzi, and D.L.Chinagila, *Appl. Spectro.* **59**, 275-279 (2005)

Mater. Res. Soc. Symp. Proc. Vol. 1134 © 2009 Materials Research Society   1134-BB08-13

Strain sensitivity in ion-implanted polymers

Giovanni Di Girolamo[1], Marcello Massaro[1], Emanuela Piscopiello[1], Emanuela Pesce[1]
Ciro Esposito[1], Leander Tapfer[1] and Marco Vittori Antisari[2]

[1] ENEA, Dept. Adv. Phys. Technol. and New Materials (FIM), Brindisi Research Center, Strada Statale "Appia" km 713, 72100 Brindisi, Italy
[2] ENEA, Dept. Adv. Phys. Technol. and New Materials (FIM), Casaccia Research Center, Via Anguillarese 301, 00123 Rome, Italy

ABSTRACT

Ion implantation process was used to fabricate ultra-thin conducting films in inert polymers and to tailor the surface electrical properties for strain gauge applications. To this aim, polycarbonate substrates were irradiated at room temperature with low energy Cu^+ ions of 60 keV at 1 $\mu A/cm^2$ and with doses ranging from $1x10^{16}$ to $1x10^{17}$ ions/cm^2. XRD and TEM measurements on the nanocomposite surfaces demonstrated the spontaneous precipitation of Cu nanocrystals at $1x10^{16}$ ions/cm^2 fluence. These nanocrystals were located at about 50 nm - 80 nm below the polymer surface in accordance with TRIM calculations. Optical absorption spectra exhibited a surface plasmon resonance (SPR) at 2 eV, in accordance with the formation of Cu nanoparticles. For doses of $5x10^{16}$ ions/cm^2 the formation of a continuous nanocrystalline Cu subsurface film occurred and a well pronounced SPR peak was observed. Otherwise, for higher doses ($1x10^{17}$ ions/cm^2) a damaged and structurally disordered film was obtained and the SPR peak was smeared out. Electrical conductivity measurements clearly indicated a reduced electrical resistance for the samples implanted with a doses up to $5x10^{16}$ ions/cm^2, whereas higher doses ($1x10^{17}$ ions/cm^2) resulted detrimental for the electrical properties, probably due to the radiation induced damage. The dependence of electrical resistance from surface load was evaluated during compression tests up to 3 MPa. A significant linear variation of the electrical resistance with the surface load was found and could be related to the changes in the spatial distribution of nanoparticles inside the copper film.

INTRODUCTION

Polymer-based composites are very attractive materials for the realization of high response pressure, strain and mechanical devices. These transducers are particularly suitable for in-service monitoring and diagnostics of the structural stability and mechanical strain of components in aeronautics, aerospace and automotive. At this purpose, metallic ion implantation is a promising technology to modify the surface properties of insulating polymers [1]. Previous works have demonstrated that low energy Cu^- ion implantation in polymers may provide the precipitation of metal nanoparticles below their surface [2,3,4]. It has been also reported that low energy Co^+ ion irradiation of viscous polymers may be optimized for synthesis of a metallic dispersion or a quasi-continuous film into the polymer surface layer [5]. Since both high energy and current density produce significant thermal effects on the exposed polymer surface, the final structure would result seriously damaged and not suitable for the desired application [6].

Thereby, all the irradiation parameters have to be carefully controlled to prevent the drawbacks mentioned above.

In this work, ion implantation process was used to modify the near-surface polymer structure in order to fabricate ultra-thin conducting films in inert polycarbonate. Indeed, the increase of implanted ion doses should promote the aggregation of nanoparticles and the formation of a continuous nanostructured film. The nanocomposite structures were characterized by using X-ray diffraction (XRD), transmission electron microscopy (TEM) and optical absorption spectroscopy. Electrical conductivity of implanted layers was measured at room temperature in terms of electrical resistance, as well the corresponded dependence from surface load was evaluated during compression tests.

EXPERIMENT

Transparent amorphous polycarbonate targets (3 mm thick) were irradiated with low energy Cu^+ ions at room temperature by using a Danfisik 1090 ion implanter. The ion-beam energy and the current density of irradiation were 60 keV and 1 $\mu A/cm^2$ respectively, whereas Cu^+ ion fluence varied in the range from $1x10^{16}$ to $1x10^{17}$ ions/cm^2. The surface charging was limited by employing conductive surface masks. Figure 1 shows any Cu^+ implanted polycarbonate substrates with different doses.

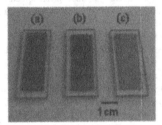

Figure 1. Photographs of polycarbonate slices implanted with Cu^+ ions with different doses: (a) $1x10^{16}$, (a) $5x10^{16}$ and (c) $1x10^{17}$ ions/cm^2, respectively.

The phase composition of nanocomposite surfaces was investigated by means of X-Ray diffraction measurements (XRD Philips PW1880) at constant incidence angle of 1° (2θ scan) and employing CuK$_\alpha$ radiation (λ=0.154056 nm). Cross sectional transmission electron microscopy (TEM FEI TECNAI F30 300 keV) was used to analyze the microstructure and morphology of the Cu implanted region. The specimens were slice cut with an ultra-microtome at room temperature.

Both optical transmittance and reflectivity of nanocomposite surfaces were measured in air in the range from 200 nm to 2500 nm by using a Varian Cary 5E spectrophotometer. The optical absorption was then calculated and correlated to photon energy.

The electrical conductivity of nanocomposite layers was measured at room temperature in terms of electrical resistance (Ω) using a Hewlett Packard 34401A digital multimeter and realizing easy electrical contacts on irradiated polymer surface. A specific measurement setup was arranged in order to evaluate the dependence of electrical resistance on surface load. In particular, several compression tests were carried out up to 2 MPa, by using a MTS Alliace RT50 tensile machine equipped with a 2 kN load cell.

RESULTS AND DISCUSSION

Structural analysis

Polycarbonate has a higher radiation resistance if compared with other polymers, thus promoting metal precipitation of copper ions. Ion implantation at different Cu ion doses could produce the precipitation of isolated nanoparticles or the formation of an embedded ultra-thin metallic subsurface Cu film in the inert polycarbonate. In Figure 2 the glancing-incidence X-ray diffraction patterns before and after ion implantation ($1x10^{17}$ ions/cm^2) are shown. The difference curve exhibits the (111), (200) and (220) cubic Cu Bragg peaks demonstrating the formation of a polycrystalline nanostructured film into the amorphous polycarbonate, in accordance with the ICDD (card no. 851326; JCPDS – ICDD2000).

Figure 3 shows the optical absorption spectra of Cu$^+$ implanted polycarbonate surfaces at different doses. With increasing photon energy, the absorbance increases and then tends to saturate. In the visible range (2-3 eV) a monotonic trend can be appreciated and explained by gradual degradation of polymer under irradiation.

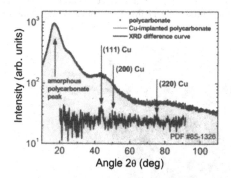

Figure 2. Glancing incidence X-ray pattern of un-implanted and Cu$^+$ implanted polycarbonate.

A typical surface plasmon resonance (SPR) was detected at photon energy of about 2 keV, suggesting the formation of metallic Cu nanoparticles or nanofilms. This finding is in agreement with other experimental results reported in literature [7].

The SPR peak is more pronounced for doses of $5x10^{16}$ ions/cm^2, whereas at higher doses is smeared out, probably due to the exposed surface damage. Indeed, ion implantation provides chemical chain scission in amorphous polycarbonate layer and with increasing ion fluence may induce a significant amount of structural defects and cracks. In addition, eventual ablation and sputtering phenomena may enhance surface damage. However, these effects should be more pronounced for high energy ion implantation [8].

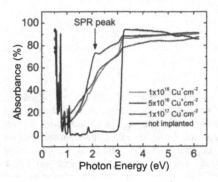

Figure 3. Optical absorption spectra of Cu^+ implanted polymers.

In Figure 4 a XTEM image of the implanted region exhibiting the nanocrystalline Cu subsurface film is shown. Cu nanocrystals (Ø~10nm) are spontaneously formed at ion doses of 1×10^{16} ions/cm^2 without further post-processing of irradiated polymer surfaces. The Cu nanocrystals are located at 50 nm - 80 nm below the polymer surface in accordance with theoretical calculations performed using the simulation code TRIM. The projected range was 75 nm, whereas the straggling was 20 nm.

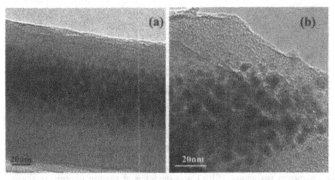

Figure 4. Cross section XTEM of implanted region, showing Cu nanofilm (a). At higher magnification (b) the precipitated Cu nanoparticles (Ø~10nm) are well visible.

The detail of Figure 4 (b) denotes the Cu nanoparticles morphology and shape. Their spatial distribution reflects the atomic profile of Cu ions in polycarbonate. At higher ion doses, *i.e.* 5×10^{16} ions/cm^2 and more, the interparticle distance results considerably reduced and the aggregation of nanoparticles produces a thin almost continuous nanocrystalline film. The lattice fringes are well observed in this embedded copper film. Otherwise, for doses of 1×10^{17} ions/cm^2, a damaged and structurally disordered film is formed, since radiation induced defects become noticeably. The nanoparticle formation mechanism is in agreement with the results reported by Boldyrieva *et al.* [4] for negative copper (Cu⁻) ion implantation in polycarbonate thin substrates. However, the current investigation demonstrates that a controlled nanoparticle precipitation was also promoted by Cu^+ ion implantation, without producing significant structural defects or damage, and required lower ion fluences than those reported in ref. [4].

Electrical characterization of strain gauge

As demonstrated by TEM observations, at lower dose the Cu nanoparticles, precipitated below polymer surface, result isolated and no electrical conductivity could be measured. Otherwise, a reduced electrical resistance was detected for polycarbonate targets implanted with doses up to 5×10^{16} ions/cm². Finally, higher doses (1×10^{17} ions/cm²) resulted detrimental for the electrical properties, probably due to the radiation induced damage.

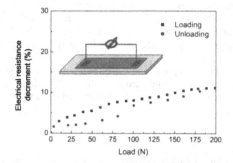

Figure 5. Electrical resistance decrement in percentange vs surface load.

Under a uniaxial compression load a significant linear variation of the electrical resistance with the surface load was found and was related to the changes in the spatial distribution of Cu nanoparticles embedded within the subsurface film. Therefore, as the load increases the electrical resistance decreases. Furthermore, only a very small hysteresis effect could be noticed during the measurements. It should be noticed that strain gauge response was strongly influenced by Cu subsurface film area, since this last one determined the starting value of its corresponded electrical conductivity. However, despite this experimental evidenced "surface area – electrical conductivity" correlation, the above reported linear trend was found for all the Cu implanted polycarbonate targets. Figure 5 shows the percentage variation of the electrical resistance with increasing surface load. The variation depends on the surface area and is typically in the range between 5 and 20 % (about 11 % in the picture) for polymer substrates (12 mm x 6 mm) implanted with a dose equal to 5×10^{16} ions/cm². Therefore, low energy ion implantation is able to produce polymer-based nanocomposites, that are particularly suitable for that development and fabrication of advanced strain gauges characterized by both high reliability and sensitivity. It should be also specified that the subsurface nanocrystalline Cu films remained stable in time without oxidizing and preserving the characteristic mechanical and electrical properties.

CONCLUSIONS

Copper nanocrystals or nanocrystalline films, *i.e.* subsurface nanostructures, were fabricated by ion implantation in amorphous polycarbonate substrates. The structural properties of ultra-thin nanocomposite layers were investigated for different process parameters. TEM, XRD and optical absorption measurements proved that at low ion doses ($1x10^{16}$ ions/cm^2) dispersed and isolated nanocrystals were formed, whereas at $5x10^{16}$ ions/cm^2 a thin continuous polycrystalline film was produced. For higher doses, a damaged and structurally disordered film was detected. The nanocomposite layers implanted at $5x10^{16}$ ions/cm^2 showed a reduced electrical resistance and a very promising strain sensitivity. To the purpose, a significant linear variation of the electrical resistance with the surface load was found and could be related to the changes in the spatial distribution of Cu nanoparticles into the subsurface film. Furthermore, only a very small hysteresis effect could be noticed. Therefore, conductive ultra-thin nanocomposite layers, *i.e* metal films embedded in polymers, are structures suitable for fabrication of innovative strain gauges (on micro- and nanoscale) and may be easily incorporated in more complex electronic devices.

Current experimental works are focused on the optimization architecture of strain gauges and their in-service response. Thereby, further studies are conducted using different typologies of targets and dopants, since irradiation parameters surely play a fundamental role on the structural, optical and electrical properties of polymer layers.

ACKNOWLEDGEMENTS

The authors would like to thank F. Valentino for his valuable technical assistance in mechanical testings. This work is supported by the Regione Puglia (Bari, Italy) – Project PONAMAT (PS_016).

REFERENCES

[1] N. Kishimoto, M. Umeda, Y. Takeda, V.T. Gritsyna, T.J. Renk and M.O. Thompson, Vacuum 58, 60 (2000).
[2] N. Umeda, V.V Bandourko,V.N. Vasilets, N. Kishimoto, Nucl.Instr. and Meth. B 206, 657 (2003).
[3] H. Boldyrieva, N. Umeda, O.A. Plaksin, Y. Takeda and N. Kishimoto, Surf.Coat.Technol. 196, 373 (2005).
[4] H. Boldyrieva, N. Kishimoto, N. Umeda, K.Kono, O.A. Plaksin and Y. Takeda, Nucl. Instr. and Meth. B 219-220, 953 (2004).
[5] S. N. Abdullin, A. L. Stepanov, Y. N. Osin, R. I. Khaibullin and I. B. Khaibullin, Surf. Coat. Technol. 106, 214 (1998).
[6] M. Ikeyama, Y. Hayakawa, M. Tazawa, S. Nakao, K. Saitoh, Y. Miyagawa and S. Miyagawa, Thin Solid Films 281-282, 529 (1996).
[7] E.K.Williams, D. Ila, A. Darwish, D.B. Poker, S.S. Sarkisov, M.J. Curley, J.C. Wang, V.L. Svetchnikov and H.W. Zandbergen, Nucl. Instr. and Meth. B 148, 1074 (1999).
[8] S.Bouffard, B. Gervais and C. Leroy, Nucl.Instr. and Meth. B 105, 1 (1995).

Mater. Res. Soc. Symp. Proc. Vol. 1134 © 2009 Materials Research Society 1134-BB08-14

A New Type of Display Device Based on Remote Swelling and Collapse of a pH-Responsive Microgel

Joseph P. Cook and D. Jason Riley
Department of Materials, Imperial College London, Exhibition Road, London, SW7 2AZ, U.K.

ABSTRACT

Microgel particles of poly(2-vinylpyridine) cross-linked with 0.5 wt% divinylbenzene were synthesised. Due to differences in the scattering of light, a dispersion of the particles appears opaque milky white above pH 4.5 and transparent below pH 4.5. The pH of a 0.15 wt% dispersion was modified using a poly(aniline) film as a source/sink of protons and a Ag/AgCl electrode as a source/sink of chloride counterions. The absorption of the sample at 400 nm was typically reduced from *c.a.* 3.5 to *c.a.* 0.5 in less than 20 minutes and increased to *c.a.* 3.2 in less than 20 minutes. This system could form the basis for a new type of display device, but it is insufficiently fast or consistent in its current form.

INTRODUCTION

A microgel particle is a colloidally sized cross-linked polymer particle that can respond to an external stimulus such as pH, temperature or ionic strength [1]. The most widely studied microgels are based on N-isopropylacrylamide, but other common monomers are acrylic acid, methacrylic acid and vinylpyridine, which impart pH-sensitivity. Concentrated dispersions of microgel particles can exhibit a vast difference in turbidity between the collapsed and swollen states. This arises because both the size and composition of the particles is different in these two states. Swollen particles have a high solvent content and so provide a lower degree of contrast with the surrounding solvent than the collapsed particles and as a result they scatter light less efficiently. Strong scattering from the collapsed particles results in a milky white dispersion and weak scattering from the swollen particles results in a more transparent dispersion. This difference in turbidity could form the basis for a new type of display device, where partitioned sections containing dispersions of the particles are placed in front of a black background and different sections are set in the collapsed and swollen states to produce an image. In order for this to be realised, a remote method would be required for switching between the two states. Potential benefits of such a device are similar to those of electrophoretic displays: there is no energy input requirement to maintain a static image and the mechanical flexibility of the device is not limited by the microgel. As such, it is likely that microgels would be most suited to devices such as electronic paper as is the case for electrophoretic displays.

Sawai *et al.* [2] achieved remote changes in turbidity of a dispersion of a microgel consisting of acrylic acid and methylmethacrylate cross-linked with ethyleneglycol dimethacrylate by cycling the potential of a poly(pyrrole) coated platinum mesh electrode. Poly(pyrrole) releases or incorporates hydroxide ions depending on the potential applied to the electrode and so allows control over the solution pH. Sawai *et al.* used this to modulate the pH of a dispersion of their acidic microgel, resulting in modulation of the particle size and thus the turbidity. The range of the transmission change achieved was extremely small at only around 1% between 90-91%, however, and this would be of no use in a real device. The main drawback of the system of Sawai *et al.* was that the microgel used varied in size by a relatively small amount: dynamic light scattering measurements revealed collapsed diameters of around 80 nm and swollen diameters of around 140 nm thus explaining the small modulation in transmission obtained. The poly(2-vinylpyridine) (PVP) microgel particles used in this study vary in diameter by a much larger ratio, from collapsed diameters of around 140 nm to swollen diameters of around 780 nm. In addition there is a greater difference in solvent content and hence refractive index between the two states for the PVP system because there is no polar spacer present, such as the methylmethacrylate in the particles of Sawai et al.. As a result a much larger change in turbidity can be achieved. The poly(pyrrole) redox couple used by Sawai et al. can only be used to modify pH for basic solutions because it operates by incorporation and release of hydroxide ions. This system is of no use with PVP as the swelling transition occurs at a pH of around 4.5. However, other conductive polymers have pH-dependent redox couples and some of these involve incorporation and release of protons rather than hydroxide ions, for example poly(aniline) (PANI) [3].

EXPERIMENT

Microgel Synthesis

The synthesis of 0.5 wt% cross-linked PVP microgel particles has been described previously [4]. 494 g of deionised water (Triple Red, > 18 MΩ cm) was placed in a 1 dm^3 round-bottomed flask, degassed with Argon (Pureshield, BOC) and heated to 72 $\pm$ 1 °C. This was stirred continuously at 300 rpm with an overhead stirrer. 4.9750 g of 2-vinylpyridine (Fluka) and 0.0455 of divinylbenzene cross-linker (55%, Aldrich) were added, followed 5 minutes later by 0.2000 g of V50 initiator (2,2'-azobis(2-amidinopropane) dihydrochloride, Aldrich) in 0.78 g of deionised water. After 5 hours the dispersion was placed into visking tubing (Medicell International, size 5) and dialysed against 15 dm^3 of deionised water for 14 days with two changes of water per day. The conversion of monomer to polymer was estimated gravimetrically by drying a known mass of the dispersion and measuring the dry mass.

Synthesis of PANI Films

Electrochemical synthesis of PANI was carried out with reference to the procedure of Greef et al. [5]. Platinum mesh working and counter electrodes (100 mesh woven from 0.0762 mm diameter, 25 mm × 25 mm, Alfa Aesar) and a Ag/AgCl reference electrode (CH Instruments) were placed in an electrolyte, consisting of 50 cm^3 of an aqueous solution of 0.1 M aniline (BDH) and 1 M sulfuric acid (BDH). The potential was cycled with a potentiostat (Autolab) between –0.155 V and +0.845 V vs. Ag/AgCl for the first 10-20 cycles after which the anodic limit was reduced to +0.745 V vs. Ag/AgCl. A total of up to 1500 cycles were undertaken, all at a scan rate of 50 mV s^{-1}. The resulting thick film of PANI was washed with deionised water several times before use.

Manipulation of pH with PANI

The PANI-coated platinum mesh electrode was used as the working electrode with a Ag/AgCl reference electrode (CH Instruments) and a Ag/AgCl counter electrode (prepared in-house [6]). The working electrode was held at set potentials of +0.8 V and -0.2 V vs.

Ag/AgCl with a potentiostat (Autolab). A small glass cylindrical sample vial (Samco, 25 mm diameter × 50 mm height) was used with the minimum volume that would cover the electrodes, 10 cm³. Samples consisted of either distilled water or PVP microgel, both of which were adjusted to the required initial pH by adding 0.1 M hydrochloric acid (Fisher). The solution was stirred at around 100 rpm with a magnetic stirrer. During potential control the pH of the sample was measured periodically with a pH meter (Orion 2 star, Thermo Scientific). In the case of the microgel samples the absorbance was measured with an absorption spectrometer (Perkin Elmer Lambda Bio 10).

DISCUSSION

Around 200 μl of 0.1M hydrochloric acid is required for a significant change in opacity for a 4 cm³ sample of the dispersion at the as-prepared concentration of 0.75 wt%. Obviously a dispersion of lower concentration would require a smaller amount of acid, and this would make remote swelling easier to achieve as a smaller amount of charge would be needed. Dilution will reduce the turbidity, however, and too low a concentration will have insufficient contrast between the swollen and collapsed states. Figure 1 shows the variation in turbidity with concentration. There is little decrease in turbidity for the collapsed state even if the gel is diluted by up to five times the as-prepared concentration from 0.75% to 0.15%, with the absorbance remaining above 3 for all visible wavelengths. Further dilution results in larger decreases in absorbance, particularly at high wavelength. The decrease in turbidity of the swollen gel is gradual over the entire range of concentrations used and this decrease is beneficial to the contrast between the two states. The samples used were diluted by a factor of five to 0.15% as this provides particularly good contrast between the collapsed and swollen states and requires significantly fewer protons to be added or removed to induce the transition between the states than for the as-prepared dispersion.

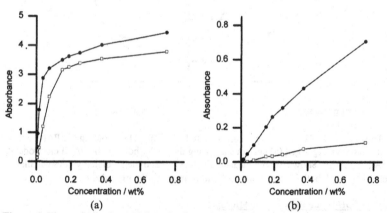

(a)   (b)

Figure 1. The variation in turbidity of a dispersion of 0.5% cross-linked PVP microgel particles at a series of levels of dilution from the as-prepared concentration of 0.75% in a cuvette of path length 1 mm, when collapsed at pH 7 (a) and when swollen at pH 3 (b). The data shown are for wavelengths of 400 nm (●) and 700 nm (□).

Modification of pH in the Absence of Microgel

Figure 2 shows the change in pH obtained in the absence of microgel by holding a PANI film at +0.8 V vs. Ag/AgCl to reduce the pH and -0.2 V vs. Ag/AgCl to increase the pH. There is a marked difference between the measured pH change and the expected pH change, calculated from the charge passed assuming that all charge passed results in proton uptake or release. The amount of charge required to bring about a desired pH change is much larger than calculated, particularly between pH 4 and 5. This is a problem because pH 4-5 is the buffering region for PVP, *i.e.* the pH range of interest in the envisaged device. The excess charge required could result from buffering of the PANI film, but the pK_a of emeraldine base PANI is not well established. As the solutions in this investigation did not contain an ionic buffer the pK_a of the films might be expected to lie in the range ~2.5-3.0, in line with the results of other authors not using an ionic buffer [7], but this would be likely to result in poor conductivity above pH 3.0 as most of the PANI would exist in the insulating emeraldine base form and this was not observed.

In some instances, increasing the pH using PANI resulted in a grey precipitate, which sedimented at the bottom of the vial. No precipitate was observed in any experiment in which the pH was reduced. This was believed to be a problem with nucleation of chloride ions onto the Ag/AgCl counter electrode and so the length of the silver wire used for the counter electrode was increased from under 100 mm to at least 200 mm. This eliminated the appearance of the precipitate. It is thus likely that the precipitate was indeed AgCl resulting from poor nucleation.

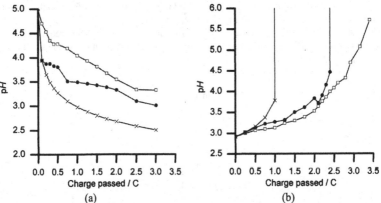

(a)   (b)

Figure 2. The change in pH with charge passed for reducing the pH by holding the PANI electrode at +0.8 V vs. Ag/AgCl (a) and for increasing the pH by holding the PANI electrode at -0.2 V vs. Ag/AgCl (b). Shown are the measured values (□) and the predicted values calculated from the charge passed, both incrementally (●) and from the initial pH value (×).

Modification of pH in the Presence of Microgel

Holding the potential of a PANI film at +0.8 V vs. Ag/AgCl for 20 minutes in a 0.15 wt% PVP microgel dispersion resulted in a reduction in pH that was sufficient to induce microgel swelling, changing the dispersion from opaque to transparent. Holding the potential at -0.2 V vs.

Ag/AgCl for 20 minutes resulted in a subsequent increase in pH that was sufficient to induce microgel collapse, returning the dispersion back to opaque. A quantity of charge between 2-3 C was found to be sufficient to induce swelling and collapse and in some cases significantly less charge resulted in a noticeable difference in turbidity, in reasonable agreement with the calculated quantity despite the possible buffering effect of the PANI film. The pH remained between 4 and 5 during potential control in the presence of the microgel. Figure 3 shows the absorption spectra of the microgel dispersion before potential control and after remote swelling and after subsequent remote collapse.

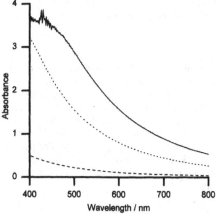

Figure 3. The typical change in turbidity of a 0.15 wt% dispersion of PVP microgel particles from the initial spectrum (solid line), after remote swelling (dashed line) and after remote swelling and remote collapse (dotted line).

Although remote swelling and collapse has been achieved, there are many ways in which the system could be improved. In particular, the change in turbidity requires at least 10 minutes and in most cases significantly longer, much too slow for a practical display device. The success of remote swelling and collapse also remains inconsistent. The rate of response is sensitive to the thickness of the PANI and the extent to which it has previously been used. Only with a thick, freshly prepared PANI film is remote swelling and collapse guaranteed to be successful in 20 minutes or less. Films that are thinner or that have been used a number of times pass charge less quickly and so require longer to induce swelling or collapse. Decreasing the sample volume would be likely to improve the success rate as the buffering effect would be reduced. This would reduce the response time and increase the useful lifetime of the PANI as less charge would need to be passed. The experiment would need modification, however, as there would be insufficient space for the current electrode arrangement. A working electrode of larger surface area would also be beneficial, as thinner films could be used which behave in a more consistent manner.

Whilst the concept of a remote swelling system has been demonstrated, a real display device requires further work. Among the factors worthy of extensive additional consideration are the choice of microgel and the choice of pH control system. Given the success in rapid swelling and collapse obtained by Armes and coworkers [8] with PVP particles stabilised with poly(ethylene glycol), remote swelling and collapse could be attempted with similar particles. Other microgels could also be investigated, for example poly(methacrylic acid) (PMAA). A major advantage of using particles based on PMAA is that the pK_a is close to 7 and so fewer protons would need to be taken up or released to induce swelling or collapse, by a factor of several orders of magnitude. Other conductive polymers could be investigated, such as derivatives of PANI and poly(pyrrole). A two electrode system may also be more practical for modifying pH on a smaller scale. Concerning device design, Sawai *et al.* [9] proposed an arrangement for a display device based on pH controlled flocculation of microgel particles that could be applied to any system based on pH-dependent scattering. Their proposal consisted of a grid of cells containing the dispersion, with one vertical side of each cell coated with the active material, which was poly(pyrrole) in their case, and another side formed the counter electrode.

CONCLUSIONS

Remote swelling and collapse of a PVP microgel dispersion has been achieved using a poly(aniline) film as a source/sink of protons and a Ag/AgCl electrode as a source/sink of chloride counterions, resulting in large changes in the turbidity of the dispersion; typically the absorption of the sample at 400 nm was reduced from *c.a.* 3.5 to *c.a.* 0.5 in less than 20 minutes and then increased to *c.a.* 3.2 in less than 20 minutes. The system is reproducible, but is sensitive to the condition of the poly(aniline) film, in particular to its age and extent of previous use. Although it has been shown that remotely modifying the scattering of a suspended particle such as a microgel could form the basis for a display device, a great deal of further work would be required in order to develop such a device. In particular, the system used here operates on too large a scale and is insufficiently fast or consistent.

ACKNOWLEDGMENTS

The authors would like to thank the EPSRC for funding.

REFERENCES

1. B.R. Saunders and B. Vincent, J. Chem. Soc. Faraday T. 92, 3385 (1996).
2. T. Sawai, H. Shinohara, Y. Ikariyama, M. Aizawa, J. Electroanal. Chem. 297, 399 (1991).
3. T. Sawai, H. Shinohara, Y. Ikariyama, M. Aizawa, J. Electroanal. Chem. 283, 221 (1990).
4. J.P. Cook and D.J. Riley, Adv. Colloid Interfac. (in press).
5. R. Greef, M. Kalaji, L.M. Peter, Faraday Discuss. 88, 277 (1989).
6. G.J. Janz, in *Reference Electrodes: Theory and Practice*, edited by D.J.G. Ives and G.J Janz (Academic Press, New York, 1961), p179.
7. (a) J.-C. Chiang, A.G. MacDiarmid, Synthetic Met. 13, 193 (1986). (b) D. Orata, D.A. Buttry, J. Am. Chem. Soc. 109, 3574, (1987).
8. (a) J. Yin, D. Dupin, J. Li, S.P. Armes, S. Liu, Langmuir 24, 9334 (2008). (b) D. Dupin, J. Rosselgong, S.P. Armes, A.F. Routh, Langmuir 23, 4035 (2007).
9. T. Sawai, S. Yamazaki, Y. Ishigami, Y. Ikariyama, M. Aizawa, J. Electroanal. Chem. 322, 1 (1992).

Mater. Res. Soc. Symp. Proc. Vol. 1134 © 2009 Materials Research Society   1134-BB08-17

Characterization of the Charging and Long-Term Performance of Cytop Electret Layers for MEMS Applications

U. Bartsch, J. Gaspar, and O. Paul
Microsystem Materials Laboratory, Department of Microsystems Engineering (IMTEK), University of Freiburg, Germany

ABSTRACT

This paper reports on the characterization of the charge stability of an amorphous fluoropolymer electret called Cytop. Cytop is a dissolved polymer material, compatible with standard micromachining fabrication technologies. In this study, Cytop layers are deposited and patterned on Pyrex and silicon substrates, followed by the electrical poling of the material by corona discharge using a customized charging station. The long-term performance of Cytop as an electret material is evaluated as a function of several relevant charging parameters. The results reveal highly stable layers, able to keep at least 92% of the initial charge 143 days after the corona charging stored at 23°C.

INTRODUCTION

Research in the field of polymer electrets compatible with microelectromechanical systems (MEMS) is of great interest because of the potential application of these materials in electrostatic micro energy harvesting devices [1], [2]. The characterization of the performance of these materials as electrets as well as their compatibility with standard MEMS fabrication processes are therefore necessary for their implementation in actual micro devices. The amorphous fluoropolymer Cytop CTL-809M (*Asahi Glass, Tokyo, Japan*) is available in solution, making it possible to spin-coat Cytop on standard substrates such as Pyrex and silicon (Si) wafers. Cytop is a multifunctional material, not only used as an electret material, but also for example for optical applications such as wave-guides [3]. In addition, both wafer and flip-chip bond processes using Cytop show great potential for the packaging of MEMS devices [4].

EXPERIMENTAL PART

Fabrication of samples

The process used to process the Cytop samples is schematically shown in Fig. 1. A Cr/Au/Cr multilayer with thickness values of 10 nm, 300 nm and 40 nm, respectively, is first evaporated on the front of a Si wafer. It is followed by the spin-coating of a multilayer of Cytop onto the metallization, using a rotation rate of 450 rpm during 20 sec, as shown in Fig. 1 (a). This step may be repeated in order to obtain the desired layer thickness. Layers thicker than 20 μm can be obtained in this way, if required. A softbake of 15 min at 100°C is performed after each spin-coating step. After deposition of the layers, the Cytop is hardbaked for 1h at 185°C. The metal layer is not patterned before the Cytop spin-coating. This has one main advantage: the single Cytop layers are more homogeneous and therefore lower rotational speeds may be used for producing thicker layers. As a result, the total number of spin-coating and process steps is reduced. After the photolithography illustrated in Fig. 1 (b), which uses the photoresist *AZ9260*

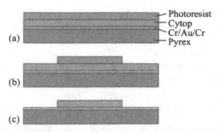

Figure 1. Fabrication of the Cytop samples: (a) Cytop and Photoresist are spun onto the metallization, (b) the photoresist is patterned by photolithography and the Cytop multilayer is structured using plasma etching.

(*MicroChemicals, Ulm, Germany*), follows the structuring of the electret shown in Fig. 1 (c). The Cytop layer is patterned using an O_2-plasma with a RF power of 100 W and a gas flow of 50 sccm using a commercial *STS Multiplex* reactive ion etching (RIE) machine (*STS, Newport, UK*). After the plasma etching the photoresist mask is removed using acetone. In contrast to other studies, no hard mask is necessary during this step as the selectivity is 4:1. The etch rate of 0.74 µm/min for Cytop is about twice as high as reported in [1]. The substrates are diced into chips with a size of 11.3×11.3 mm^2.

Corona charging

The setup to charge the electret is a typical corona discharge setup containing a corona tip, a grid electrode and a grounded sample to be charged [5], [6]. The setup is schematically shown in Fig. 2 (a). If a high negative DC voltage is applied to the corona tip, the surrounding air is ionized. The resulting ions then start to drift, crossing the grid electrode as a result of the more positive potential of the grid, to the grounded sample. The Cytop layers on single chips are charged using the custom-made corona discharge setup shown in Fig. 2 (b). The completesetup contains one high voltage (HV) source for the corona tip and one for the grid electrode. The custom-made charging station, from hereon termed "Corodis", includes the controlled heating of the samples to adjust the temperature during the charging process. The setup can be operated in constant current or constant voltage mode. Figure 2 (c) shows the schematic cross-sectional view of the "Corodis" station. It consists of a base socket, three outer conducting and isolating cylinders with the sample holder on top, and a corona tip. The distance between the tip and the grid electrode can be adjusted in the range of 1-5 cm.

The gap between the grid and the sample is also adjustable between 0.5 mm and 3.5 mm. The maximum chip size that can be charged by the "Corodis" station is 15×15 mm^2.

Measurement setup

The characterization of the resulting charge distribution in the electrets is performed using the automated measurement setup shown in Fig. 3. The system includes a vacuum chuck holder where 16 chips can be placed and sequentially characterized in a single series of measurements. This holder is mounted on two linear stages and one rotational stage with µm resolution programmed to move in a user defined sequence. An electrostatic voltmeter (*Model*

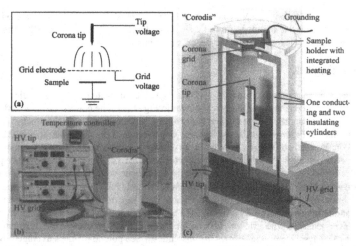

Figure 2. Corona discharge setup: (a) schematic layout, (b) View of the complete discharging setup and (c) detailed view of electret charging station.

279, Monroe Electronics) is fixed above the chuck on an independent *z*-stage that allows to adjust the measurement gap between sensor and electret surface. A measurement routine for scans was programmed that enables scans to be executed over the electret surface with a selectable step size. One-dimensional (1D) and two-dimensional (2D) scans of the electrets are possible with the setup.

RESULTS AND DISCUSSION

This section discusses the effect of various charging parameters, namely the corona grid voltage V_G, the electret charging temperature and the electret storage temperature on the long-term charge stability. All samples discussed here were corona-charged during $t_C = 5$ min using a

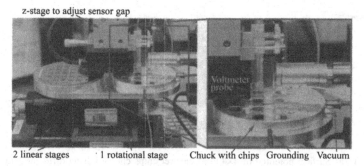

Figure 3. Automated measurement setup to characterize the charge stability and distribution of electret samples.

corona tip voltage of $V_T = -5$ kV at a charging temperature of $T_C = 23°C$, if not otherwise stated. The chips were stored at a temperature of $T_S = 23°C$ at a relative humidity of 40%. The thickness of the Cytop electret layer is $d_C = 9$ μm for all experiments.

Corona grid voltage

Typical measurement results of a 1D surface potential scan of a 4×4 mm^2 electret chip for three different grid electrode voltages are shown in Fig. 4 (a). The corona grid electrode voltage V_G is the main parameter used to control the resulting surface potential of the electret after the corona discharge. Figure 4 (b) shows the linear dependence of the maximum measured surface potential on the used corona grid voltage V_G as expected. Nevertheless, the electret surface potential does not reach the theoretically predicted value of V_G. The maximum surface potential of the electret layer is about 77% of the initial grid electrode voltage, after 1 h has elapsed since the corona-charging. Using this linear dependence, electret chips with a desired surface potential can be fabricated just by adjusting the grid voltage V_G. The effect of the applied corona grid voltage on the long-term stability of charges in Cytop is shown in Fig. 5. The surface potential of electret layers charged with grid voltages ranging from −200 V to −900 V is investigated of a period of 143 days. Figure 5 (a) shows the absolute surface potential values averaged over the electret size of 4×4 mm^2. No so-called cross-over effect, which designates the faster discharging of samples charged to higher potentials is observed. Nevertheless, the tendency of the slower discharging of samples charged to lower surface potentials can be seen in Fig. 5 (b). The three samples charged with $V_G = -200$ V to −400 V show the lowest charge decay.

Electret charging temperature

The effect of the charging temperature T_C of the electret layer of Cytop has rarely been discussed in the literature. Nevertheless, studies concerning the performance of Cytop as an electret material suggest that applying a charging temperature T_C higher than the glass transition temperature of Cytop is advantageous for the charge stability [1], [2]. The glass transition temperature for Cytop is 108°C [6]. To verify this suggestion, Cytop samples were charged at four different charge temperatures, namely 23°C, 60°C, 90°C, and 120°C. The measurement results of the electret surface potential for the different temperatures are shown in Fig. 6. In the

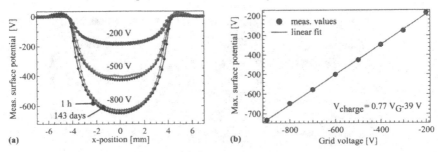

Figure 4. (a) Surface potential of 4×4 mm^2 electret chips charged using different grid voltages measured 1 h and 143 days after the corona discharge. (b) Linear dependence of the maximum surface potential and the used grid electrode voltage.

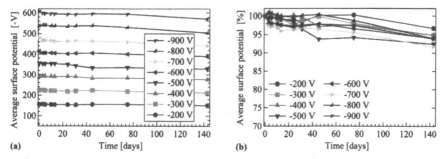

Figure 5. Long-term measurement of 4×4 mm² electret chips charged using different grid voltages. (a) Raw values of the average surface potential, (b) normalized average surface potential during the first 143 days after corona discharging.

first 143 days after corona charging, the electret temperature during the corona discharge T_C has no visible effect on the charge stability. Only a minor effect of the charging temperature of Cytop on the resulting surface potential can be seen, which leads to slightly higher values when applying a lower charging temperature.

Electret storage temperature

The immunity of the charges in an electret layer against heat is an important issue. The electret processing steps subsequent to the corona charging and the environmental conditions of the later application might demand a sufficient temperature resistance of the charges trapped in the Cytop layer. Therefore, the effect of the storage temperature of charged Cytop was investigated. The measured temperature induced discharge of the Cytop layer at 100°C, 120°C, and 140°C is shown in Fig. 7. For each temperature the charge decay of 4 samples was tested. The storage temperature of the corona-charged Cytop has little effect on the charge stability of Cytop during the first 10 h after the corona discharge. At 120°C, only 75% of the initial surface potential is present 10 h after charging the electret. However, the increase of the storage temperature to 140°C leads to a faster decay of trapped charges. The measured graph of Fig. 7 is in good agreement with the temperature stimulated discharge (TSD) of Cytop presented in Ref.

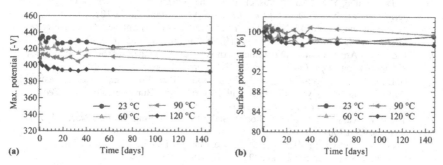

Figure 6. Effect of the electret charging temperature T_C on the long-term charge stability.

227

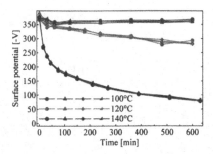

Figure 7. Temperature induced discharge of Cytop samples. For each temperature (100°C, 120°C and 140°C) the charge decay of four samples is reported.

[2], where the discharge current of an electret material is measured as a function of the sample temperature.

CONCLUSIONS

The fabrication of samples, corona charging, and characterization of the charge stability of the amorphous fluoropolymer electret material Cytop was presented in this paper. The material can be spin-coated and patterned using standard MEMS fabrication processes. The long-term charge stability is an important issue and therefore, the effect of the corona grid voltage, the electret charging temperature and the electret storage temperature on the charge stability was measured.

ACKNOWLEDGMENTS

The financial support of the "Deutsche Forschungs Gemeinschaft" DFG under program GR1322/1 "Micro Energy Harvesting" is gratefully acknowledged.

REFERENCES

[1]   T. Tsutsumino, Y. Suzuki, N. Kasagi, Y. Sakane: Seismic power generator using high-performance polymer electret, *Proc. IEEE MEMS'06*, 2006, pp. 98-101.

[2]   T. Tsutsumino, Y. Suzuki, N. Kasagi, Y. Tsurumi, High-performance Polymer Electret for Micro Seismic Generator, *Proc. PowerMEMS Conference*, 2005, pp.9-12.

[3]   J. Zubia, J. Arrue: Plastic optical fibers: An introduction to their technological processes and applications, *Opt. Fiber Technol.* 7, 2001, pp. 101-140.

[4]   U. Bartsch, T. Huesgen, J. Gaspar, P. Woias, and O. Paul, Low-temperature Adhesive Flip-chip Bonding using Cytop combined with Electrical Stud Bump Interconnects, *Tech. Digest MicroMechanics Europe Workshop 2008*, pp. 411-414.

[5]   Y. Fei, Z. Xu, C. Chen, Charge Storage Stability of SiO_2 Film Electret, Proc. IEEE SoutheastCon 2001, pp. 1-7.

[6]   T. Oda, Surface Charge Behavior of Corona-Charged Thin Polymer Films, Proc. 12th International Sym. on Electrets, 2005, pp. 188-191.

[7]   CYTOP Amorphous Fluoropolymer, *Technical Data 2004*, Asahi Glass Co., Ltd, Japan.

Mater. Res. Soc. Symp. Proc. Vol. 1134 © 2009 Materials Research Society   1134-BB08-18

The melt electrospinning of polycaprolactone (PCL) ultrafine fibers

Chitrabala Subramanian, Samuel C. Ugbolue*, Steven B.Warner and Prabir K.Patra[+]

* Department of Materials and Textiles, College of Engineering, University of Massachusetts Dartmouth, MA 02747, USA

[+] Department of Mechanical Engineering and Materials Science (MEMS), Rice University, 6100 Main Street, Houston, TX 77005, USA

ABSTRACT:

Electrospinning is a technique of producing nanofibers from polymer solution/melt solely under the influence of electrostatic forces. In this research, we investigated the formation of nanofibers by melt electrospinning polycaprolactone (PCL). The effect of process parameters such as molecular weight, applied voltage, and electrode separation on the fiber diameter was investigated. Controlling the process parameters could help increase the proportion of ultrafine fibers in the melt electrospun nonwoven mat. The velocity of the straight jets was in the range of 0.2-1 m/s. The melt electrospun fibers were characterized with respect to fiber diameter, distribution, mechanical properties and birefringence. Melt electrospun polycaprolactone fibers had a diameter distribution of the order of 5 -20 μm. The birefringence of the melt electrospun fibers increased with decrease in fiber diameter.

1. INTRODUCTION:

Electrospinning process, known since the 1930's is a technique of producing fine fibers from polymer solution/melt solely under the influence of electric field. In this process a high voltage source charges the polymer solution; the high applied voltage produces surface charge on the droplet and at sufficiently high voltage; the drop elongates to form a Taylor cone and is ejected as continuous stream from the capillary tip. The charged jet is subjected to bending instabilities as it moves towards the grounded collector [1]. A number of polymers have been successfully electrospun from their solutions. Comparatively fewer polymers have been melt electrospun and obtaining nanofibers from polymer melts has not been an easy task [2]. Electrostatic spinning of fibers from polymer melts was first published by Lorrando and Manley in 1981. Their work described spinning of rapidly crystallizing polymer by application of electrostatic field as against conventional methods that use high pressure to extrude the polymer melt in air [3, 4, 5]. Some of the polymers that have been melt electrospun successfully are polypropylene and poly(ethylene terephthalate) [6], polyimide, poly(ethylene naphthalate), poly(caprolactone) and polypropylene in vacuum [7], polyethylene, ethylene vinyl acetate, poly(R,R lactide) [8], block co polymer of poly(ethylene glycol) and poly(ε-caprolactone) [9]. Melt electrospinning of polypropylene by addition of salts or surfactants for charge induction has been tried with the aim of increasing the charge carrying ability that will help in the formation of fine fibers [10]. Modified apparatus designs for spinning plastic polymers at elevated temperatures have been attempted in the past [11-13].

This article focuses on melt electrospinning of polycaprolactone. The details of the melt electrospinning process and the melt electrospun fibers have been studied with the aim of optimizing the process to maximize the proportion of nanofibers in the nonwoven mat. Successful production of nanofibers from polycaprolactone melts would be advantageous, since electrospun fibers are widely used in biomedical applications.

2. EXPERIMENTAL:

Our melt electrospinning apparatus consisted of a glass syringe filled with the polymer to be spun, high voltage DC supply (model ES50P-20W from Gamma High Voltage Research), heat gun and grounded collector. The polymer used was polycaprolactone, CAPA® 6000 series from Solvay Chemicals and the molecular weights were in the range of 10,000-80,000 g/mole (MFR: 40 g/10 min-3 g/10 min). The polymer pellets in the syringe were molten by heating it with a heat gun. The flow rate could also be controlled using a syringe pump (model 200 from KD Scientific) aligned vertically to the syringe. A high voltage source connected to the syringe tip charged the polymer melt. The fibers that spun under the influence of electrostatic forces were finally collected on a grounded metal grid. The images of the melt electrospinning process were captured using Sensi Cam from Cooke Corporation at the rate of 64 frames per second. The electrospun fibers were observed under JEOL JSM 5610 high vacuum scanning electron microscope to study the fiber diameter and distribution. The tensile properties of a single thick fiber were tested by ASTM standard test method D 3822 on a CRE type (Instron) tensile testing machine at a predetermined gauge length and rate of extension. Molecular orientation determination of the sample was attempted by birefringence measurements using Leica optical microscope fitted with an interference filter of wavelength 546 nm and Senarmont compensator.

3. RESULTS AND DISCUSSION:

3.1 Melt electrospinning process:

High speed images of the electrospinning process were captured to understand the process better. The different stages of the spinning process were:
1. Initiation of the jet from the molten polymer
2. Propagation of the jet
3. Solidification of the fibers
4. Fiber collection on the grounded collector.

When polycaprolactone (PCL) melt was subjected to electrostatic forces, the hemispherical drop formed at the tip of the needle began to deform. From a hemisphere it formed a cone-like structure that is called the "Taylor cone" and at certain critical voltage a steady jet was ejected. The semi vertical angle of the cone at critical voltage was measured to be approximately 36° in accordance with previous work in this area [3, 4, 5]. A distinct straight jet and whipping zone were observed. The polymer jet initially traversed a straight path and began to spiral at a distance of around 4 cm. from the collector when the distance between the syringe tip and collector was 10 cm.

The velocity was determined by using the following equation [14]:

$$M = \pi \rho r^2 v \tag{1}$$

where, M is the mass flow per second, ρ is the polymer density, r is the radius of the fiber, v is the velocity.

The average downward velocity at which the straight jets were traveling was calculated to be in the range of 0.2-0.8 m/s from equation 1. When the distance between the needle tip and collector was 7 cm. the average velocity was 0.3 m/s; that dropped to 0.26 m/s when the distance was increased to 10 cm. The polycaprolactone melt had a tendency to retain its molten state. Hence the difference in velocity with respect to the electrode separation may be marginal in the case of polycaprolactone melt. Solidification of the polymer may restrict the effective movement of jet when the jet traveling distance increases. The fibers had a tendency to fuse, indicating that they were still in the molten stage when they reached the collector. Figure 1 shows the captured images of propagation of the jet with a distinct whipping zone.

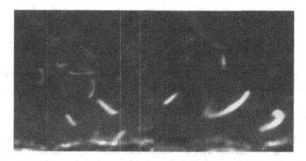

Figure 1: Images of whipping zone in melt electrospinning of PCL

3.2 Effect of process parameters:

3.2.1 Effect of molecular weight and applied electric field:

With an increase in molecular weight from 25,000 to 37,000 g/mole, the viscosity of polymer melts increased. Owing to the higher chain entanglements present in the melt, a higher force would be required by the polymer melt to spin finer fibers. The chain entanglements would resist the stretching process and isolation of the entanglements to spin fine fibers. As the molecular weight increased, a higher electric field would be required to overcome the surface tension and viscoelastic properties of the polymer melt and stretch it to form fibers [6]. Figure 2 illustrates the effect of molecular weight on fiber diameter of PCL. The applied electric field has an effect on the forces acting on the polymer drop that deform it.

A higher applied voltage provides additional force to overcome the viscoelastic forces and surface tension exerted by the polymer drop that resists the deformation. This will help in isolating the chain entanglements present in the polymer melt and eventual production of finer fibers. As the distance between the electrodes is increased, the fiber diameter dropped since it provides an additional time for the polymer jet in flight to stretch. If the polymer remains in the molten stage, increasing the distance will help in effective stretching of the fibers and hence finer fibers. Higher distance also provides additional time for the polymer in the molten state to stretch effectively which would help reduce the fiber diameter.

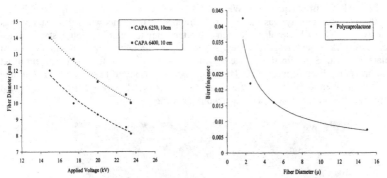

Figure 2: Effect of molecular weight and applied electric field on fiber diameter of PCL

Figure 3: Birefringence of melt electrospun PCL fibers

3.3 Birefringence measurement:

The birefringence of the fibers increased with decrease in fiber diameter as shown in Figure 3. This is in accordance with the observation made by Warner et al for melt blown polypropylene [15] and Buer et al [16] that the birefringence of electrospun fibers likely increases with decreasing diameter. Higher drawing forces would yield finer fibers and the drawing process could bring about an increase in the molecular orientation.

3.5 Mechanical Properties:

For a 15 μm fiber the initial modulus was calculated to be around 4.8 gf/den. The average elongation at break was around 190%. The stress-strain curve of unoriented fiber showed specific regions like Hookean region, yield point, neck formation, and failure. During the neck formation and propagation, area reduction takes place and these local regions propagate to the end of the specimen [17]. Also the load-extension curve showed fluctuations, which could be due to the noise from the instrument as well as necking of fibers [6].

3.6 Fiber Morphology:

Typically the fibers spun were cylindrical. Melt electrospinning of polymers also yielded different fiber morphologies, namely:

3.7.1 Fused fibers:

The fibers that arrive at the collector before being solidified would still be in the molten stage and would have a tendency to fuse to each other at the contact points of the nonwoven mat as shown in Figure 4. This kind of structure was particularly observed at higher applied voltage as the polymer jet would travel at a higher velocity towards the collector.

As a result of the short time taken by the jet to reach the grounded collector, it is possible that the polymer is still in the molten state, particularly when the separation distance for the electrodes is small. When the fiber reaches the collector, they tend to fuse to each other at the contact points.

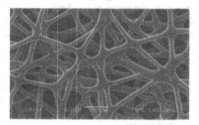

Figure 4: SEM of melt electrospun polycaprolactone fused fibers

3.7.2 Spiral fibers:

This structure was particularly noticeable amongst the finer fibers in the PCL nonwoven mat. The probable reason for the helical structures would be the whipping action during the electrospinning process. In the whipping zone, the polymer jet traverses a helical path. The jet undergoing the whipping action may have solidified, due to rapid heat transfer between the fine jet and the surrounding air, before it reached the grounded collector. Solidification being the termination step would have caused some fibers to retain their helical shape as observed in the SEM image shown in Figure 5. Thus, the helical shape of the fibers could be related to traces of instability.

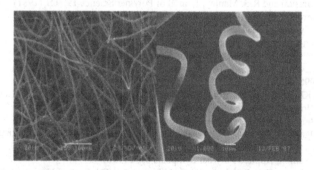

Figure5: SEM of melt electrospun polycaprolactone showing helical structure

4. CONCLUSIONS:

The process of electrospinning from melt was studied using polycaprolactone as a model polymer. Under the influence of electric field the polymer melt deformed to form a Taylor cone followed by propagation in the form of straight jet and whipping zone towards the collector. The velocity of the straight jet was in the range of 0.2-1m/s. Parameters such as applied voltage, electrode separation, molecular weight had an effect on the fiber diameter. Controlling the process parameters helped increase the proportion of fine fibers in melt electrospun polycaprolactone web but producing sub micron fibers from electrospinning of polymer melts was marginally successful. The melt electrospun polycaprolactone fibers were thicker, of the order of 5 -20 μm as against solution electrospun fibers. Finest fibers obtained by controlling the process parameters were in the range of 260 nm. Results of single fibers testing on Instron testing machine showed a mean tenacity of 1.2 g/den, initial modulus of 4.8 gf/den and extension-at-break of 190% for polycaprolactone. Birefringence of the melt electrospun fibers increased with decrease in fiber diameter.

ACKNOWLEDGEMENT

The authors are grateful for the financial support to the graduate student, Chitrabala Subramanian, through funds from the National Textile Center under the US Department of Commerce Grant 99-27-07400.

REFERENCES:

1. G.I. Taylor, Proceedings of Royal Society London, Ser A **313**, 453 (1969)
2. V.E. Kalayci, K.P. Patra, Y.K. Kim, S.C. Ugbolue and S.B. Warner, Polymer **46**, 7191 (2005)
3. L. Larrondo and R.S. Manley, Journal of Polymer Science **19**, 909 (1981)
4. L. Larrondo and R.S. Manley, Journal of Polymer Science **19**, 921 (1981)
5. L. Larrondo and R.S. Manley, Journal of Polymer Science **19**, 933 (1981)
6. J.M. Lysons, Ph.D. Thesis, Drexel University, 2004
7. R. Rangkupan, M.S. Thesis, University of Akron, 2002
8. P. Schwartz, Fiber Society, Book of Abstracts, May 2001
9. P.D. Dalton, J.L. Calvet, A. Mourran, D. Klee and M. Moller, Biotechnology Journal **1** (9), 998 (2006)
10. H.S. Khurana, P.K. Patra and S.B. Warner, Polymer Preprints **44** (2), 67 (2003)
11. Y.L. Joo and H. Zhou, U.S. Patent No. 7 083 854 (10 May 2005)
12. H. Zhou, T.B. Green and Y.L. Joo, Polymer **47**, 7497 (2006)
13. N. Ogata, N. Simada, S. Yamaguchi, K. Nakane and T. Ogihara, Journal of Applied Polymer Science **105** (3), 1127 (2007)
14. T.R. Barnett, Applied Polymer Symposia **6**, 51 (1967)
15. S.B. Warner, C.A. Perkins and A.S. Abhiraman, INDA Jour. of Nonwoven Res. **2**, 33 (1995)
16. A. Buer, S.C. Ugbolue and S.B. Warner, Textile Research Journal **71**(4), 323 (2001)
17. S.B. Warner, Fiber Science, (Prentice Hall, Inc., Englewood Cliffs, NJ, 1995), p.131

Mater. Res. Soc. Symp. Proc. Vol. 1134 © 2009 Materials Research Society 1134-BB08-27

Multi-Scale Grafted Polymeric Nanostructures Patterned Bottom-Up by Colloidal Lithography and Initiated Chemical Vapor Deposition (iCVD)

Nathan J. Trujillo, Salmaan H. Baxamusa and Karen K. Gleason
Department of Chemical Engineering, Massachusetts Institute of Technology, 77 Massachusetts Ave. 66-419, Cambridge, MA 02139, U.S.A.

ABSTRACT

Colloidal lithography is a popular, non-conventional process that uses two–dimensional self-assembled monolayer arrays of colloidal nanoparticles as masks for techniques such as etching or sputtering. Initiated Chemical Vapor Deposition (iCVD) is a surface controlled process which offers unprecedented opportunity for producing polymeric layers grafted to substrates with dangling vinyl bonds and patterned through a colloidal template. We demonstrate a generic "bottom-up" process as an inexpensive and simple technique for creating well-ordered arrays of functional patterned polymeric nanostructures. These patterns were produced from thin polymer films of p(butyl acrylate) and p(hydroxyethyl methacrylate), and are robustly tethered to the underlying substrate, as demonstrated by their ability to withstand aggressive solvents. Furthermore, using capillary force lithography, we created topographical templates for large-scale orientation of the nanoparticle assembly. Through this "top-down" approach, for assisting the bottom-up assembly, we present a process for multi-scale patterning of functional polymeric materials, without the need for expensive lithography tools.

INTRODUCTION

Materials patterning through non-conventional lithography can reduce the cost of patterning fine structures when compared to traditional nanofabrication techniques such as photolithography [1]. Liftoff techniques form patterns on a wafer surface by an additive process. A negative pattern is formed by first depositing a stencil layer, which exposes the substrate in specific locations. A film is subsequently deposited over the stencil and exposed substrate. The stencil is removed by dissolution in solvent, leaving behind a patterned film. We report on a variation of this liftoff technique; one that uses a self-assembled monolayer of colloidal particles as the stencil. Monodisperse colloidal particles can self-assemble into 2-D hexagonal arrays when deposited onto various substrates. By controlling the drying process, solution cast colloidal particles will self-assemble into a hexagonally closed packed monolayer. The ordered void array in the particle intersticies has been used as a patterning mask for over 25 years [2]. These 2-D patterns have been used to pattern carbon nanotubes [3], ZnO nanorods [4], nickel and gold dots [5, 6]. Several experimental techniques for creating monolayers using commercially available latex spheres have been described [4, 5, 7, 8]. Nanopatterns of p(acrylic acid) domes have been created from top down patterning [9], however, this method requires the use of a hydrophilic polymer and a reactive ion etch, which can destroy the delicate polymer functionality.

In 2007, bottom-up patterning of silicon dioxide was demonstrated. This study used PECVD and liftoff to create arrays of hemispherical depressions, arranged in an HCP pattern, for producing micro contact stamp molds [10]. The large hemispherical surface area and the excellent substrate adhesion make the general formulation of these patterns attractive for various applications that require functional nanostructures, such as catalysts, sensors, antifouling surfaces, nanoactuators, and photovoltaics; however, these types of patterns are yet to be synthesized by vapor phase techniques. Furthermore, PECVD, a workhorse for the semiconductor industry, is a high energy process which can destroy organic moieties through unwanted side reactions [11-14]. iCVD, a variation of hot filament chemical vapor deposition developed in our lab, is a solvent-free process [15, 16] which requires much lower energy density for deposition when compared to PECVD (0.02-0.12 W/cm^2[15] versus 0.13-2.1 W/cm^2 [16, 17]), and allows for full retention of delicate functional groups [16]. Unlike PECVD mechanisms, iCVD polymerization is well understood and most closely resembles free radical polymerization in solution [15]. Because iCVD is a surface controlled process [18] it affords unprecedented opportunity for producing patterned polymer polymeric layers grafted to surfaces with dangling vinyl bonds [16]. Vinyl groups covalently anchored to a surface can react with the initiating species by the same free radical mechanism responsible for polymerization of the vinyl monomers, only monomeric propagation now proceeds from the surface bound radical. The ability to produce films with adhesion contrast is important to the novel patterning technique which we develop in this report. We employed this simple principle of colloidal lithography, coupled with iCVD to create a generalized process for bottom-up patterning of functional polymers.

Practical implementation of these robust functional polymer arrays would require large scale order; the self-assembly must contain few point and line defects. By providing a topographically patterned template, which can also be produced via non-conventional lithography, one could achieve long range order within a nano-patterned regime [19]. Therefore, precisely defined large-scale features can drive rational design at the nanoscale [19](since the colloidal crystal template defines the lowest level for hierarchical polymer patterns). Templated self-assembly can produce spatially addressable nano-features with low line-edge roughness [19] for various device applications. We employed capillary force lithography techniques to form hierarchically patterned iCVD nanostructures, without the need for expensive lithography equipment.

The methods we present in this report are generic, and can be extended to create addressable patterns of grafted functional nanostructures from any iCVD polymer.

EXPERIMENTAL

A generic process for creating patterned polymer using self assembled colloidal arrays is shown in Figure 1. A silicon substrate (100) was treated with oxygen plasma to remove any organic impurities [20] and to increase the surface hydroxyl concentration [21]. Increasing the hydroxyl concentration not only provides more surface sites for subsequent silyation chemistry, but also increases surface hydrophilicity which is critical for forming a monolayer of the colloidal particles [22]. A solution of monodisperse 1 μm

236

poly-styrene nanoparticles, 2.5% wt. in water, was mixed 1:1 with a surfactant solution (Triton X-100:methanol/1:400 volume) [7], cast onto the plasma cleaned wafer in discrete 2.0 μL droplets, and allowed to dry under ambient conditions for 20 minutes (Figure 1a). The samples were then loaded into a vacuum oven and exposed to tricholorvinyl silane vapor. This vinyl-silane coupling agent binds covalently onto the silicon substrate through hydrolysis of the chlorine moieties by surface hydroxyls. This occurs only in the exposed regions of the substrate, through the particle interstices (Figure 1b). A blanket polymer film was then grafted onto the substrate via iCVD (Figure 1c). In this report, homopolymer patterns were created from films produced with either 2-hydroxyethyl methacrylate (HEMA) or n-butyl acrylate (BA) monomer precursors. The films were deposited in a custom-built iCVD reactor, which has been described previously[18]. The deposition parameters for each type of film were adopted from Chan and Gleason [23] and Lau and Gleason [18], respectively. Approximately 500 nm of each film was deposited. To remove the polystyrene particles and any polymer not bound to the surface, the grafted films were subsequently placed in an ultrasonic bath and rinsed in THF for 10 minutes. This step revealed largely ordered honeycomb-like arrays of polymer, which was covalently bound to the substrate through free radical polymerization from the substrate-tethered vinyl group (Figure 1d).

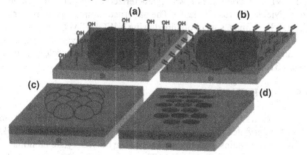

Figure 1. A generic process scheme for producing patterned polymeric nanostructures using colloidal lithography. A hydroxylated substrate, which has been treated with an oxygen plasma, serves as a hydrophilic base for depositing a 2-D assembly of colloidal nanoparticles (a). The masked sample is then treated with a silane coupling agent, which is covalently attached to the substrate in the exposed regions of the colloidal mask (b). This acts as an adhesion promoter to graft the functional iCVD polymer which is subsequently deposited (c). The grafted film is sonicated in THF to remove the colloidal template and any ungrafted polymer. This reveals an array of nanostructures patterned in a hexagonal arrangement (d).

A topographical template was created for the nanoparticle assembly using capillary force lithography [24]. A 1.5% wt. solution of polystyrene (MW$_n$ = 45,800), dissolved in toluene, was spin-coated onto a plasma-cleaned silicon wafer, to form a 300 nm thick film. A pre-patterned PDMS mold, created from a polyurethane master, was placed onto the polystyrene film and pressed at 150°C, for 2.5 hours. This forced the polystyrene to be siphoned into the grooves of the mold (Fig 2a/2b). The pressed assembly was cooled at room temperature for one hour, to allow the polystyrene to solidify; the mold was subsequently removed. The patterned silicon substrate was then exposed to oxygen plasma for 1 minute. This etched away any polystyrene that remained in the grooves. The colloidal nanoparticles were deposited into the patterned grooves (Fig 2c) and the grafted iCVD polymers were patterned, according to the methods outlined above. The lift-off step removed the patterned grooves as well as the nanoparticle assembly within the grooves, revealing hierarchal patterns of grafted iCVD polymer.

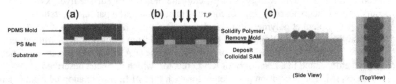

Figure 2. A generic process scheme for creating a topographical template used for producing hierarchical patterns of polymeric nanostructures, by capillary force lithography. A patterned PDMS mold is pressed onto a polystyrene melt (a/b). After cooling, the mold is removed. A SAM of colloidal particles is deposited into the grooves (c). This hierarchical template is used for patterning the iCVD polymers, according to Figure 1.

The iCVD film composition was characterized by transmission mode FTIR spectroscopy (Nicolet Nexus 870 ESP). The patterned samples were sputter coated with 5 nm of gold and SEM images were obtained using a JEOL JSM 6060 with 5 kV acceleration voltage. Optical microscope images were obtained by Olympus CX-41 microscope and AFM images were generated using a Veeco Metrology Nanoscope IV/Dimension 3100 Scanning Probe Microscope in tapping mode.

RESULTS AND DISCUSSION

Figure 3 shows FTIR spectra for iCVD films produced from HEMA and BA precursors. Alkyl acrylates have a characteristic carbonyl stretch between 1732 and 1736 cm^{-1} [18]. This peak remains unchanged between the butyl acrylate monomeric precursor (Figure 3a) and polymer film (Figure 3b). This demonstrates the retention of the monomer's functional groups after polymerization. Furthermore, the poly (butyl acrylate) polymer can be identified through the double band at 950 cm^{-1} [18] and by the absence of the vinyl peaks present in the monomer precursor at at 1410 cm^{-1} and between 1630 cm^{-1}-1650 cm^{-1}. HEMA contains five main vibrational modes [23]. The C=O stretching modes between 1750-1690 cm^{-1}, C-O stretching (1300-1200 cm^{-1}), and C-H bending at 1500-1350 cm^{-1} correspond to both the spectra obtained for the monomer precursor (Figure 3c) and as-deposited film (Figure 3d). There is also a broad peak centered at 3450 cm-1 (not shown) which is present in both monomer and polymer spectra. This signifies that the functional hydroxyl group is retained after polymerization. This is consistent with the results reported by Chan and Gleason [23] which indicated that the entire functional pendant hydroxyethyl group is preserved in the iCVD polymer. P(HEMA) behaves as a hydrogel and swells upon water absorption [23]; thus, by preserving the delicate functionality, iCVD preserves the ability to create patterned, responsive surfaces.

238

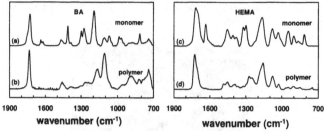

Figure 3. FTIR spectra for (a)butyl acrylate monomer, (b) corresponding iCVD film, (c) HEMA monomer, (d) corresponding iCVD film. The spectral similarity between monomers and their corresponding iCVD film suggests the retention of functional groups in the films.

Figure 4a shows an optical microscope image of a cast monolayer of 1μm polystyrene in a 2-D hexagonal lattice. The above experimental procedure created regions of continuous monolayer coverage with an area greater than 1,000,000 μm². This method was not optimal for creating the large monolayer coverage; however, techniques have been described [4] for depositing a monolayer of colloidal particles over areas larger than 1 cm². Figure 4b contains an SEM image of p(butyl acrylate) patterned with 1 μm spheres. This image shows a large area of the honeycomb pattern that was produced after the colloidal template and ungrafted homopolymer was removed with THF. The image inset is an enlargement showing the regularity of the patterns, with the smallest features about 150 nm in width. The same results were observed with the patterns produced from HEMA (Figure 4c). This SEM image clearly shows the interface between the colloidal monolayer and the silicon substrate. The patterns extend to the outer-most portion of the monolayer and fully capture the geometry of the particles in the periphery.

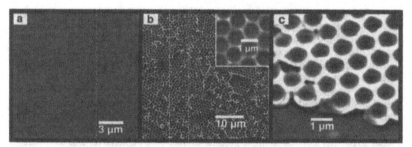

Figure 4. An optical microsope image for a 2-D assembly of 1 μm colloidal particles (a). SEM images for (b) patterned p(butyl acrylate)and (c) p(HEMA). The expanded image (inset) for butyl acrylate shows the well- ordered grafted features which are as small as 150 nm in width.

The butyl acrylate pattern preserves the hemispherical geometry of the colloidal template. The AFM image in Figure 5 shows that the particles were successfully removed via liftoff by ultra sonication in THF and that the geometry of the polymer wells, with features as tall as 500 nm, preserves the hemispherical HCP lattice generated by the self-assembled monolayer template.

Figure 5. Tapping mode AFM image for patterned p(butyl acrylate) "nanobowls". The polymer pattern preserves the hemispherical geometry from the colloidal template.

It is crucial to emphasize the importance of the silyation step in the above process; without adding the adhesion promoter one will not retain the polymer after the sonication step. The homopolymers of both p(butyl acrylate) and p(HEMA) are extremely soluble in common solvents; however, the patterns in Figure 4 were imaged after 10 minutes of sonication in THF and additional overnight soaking. Therefore, the covalent attachment between the polymer and the substrate provides a robust interface which renders the patterns insoluble in aggressive solvents.

Figure 6a is an optical micrograph that depicts a set of polystyrene lines, patterned by capillary force lithography. The PDMS mold contained sets of grooves of varying width, with a 6 μm pitch. Therefore, the patterns transferred onto the polystyrene were capable of confining a monolayer of six, close- packed 1 μm polystyrene particles (Figure 6b). This hierarchical assembly was then used as a template for grafting iCVD polymer. After the template was lifted off, hierarchically patterned, grafted iCVD polymer was revealed. These patterns contain two periodicities, at different length scales. The first periodicity, corresponding to the larger length scale, is attributed to the polystyrene template. The iCVD patterns are separated from one and other by a distance equal to the width of the grooves (Figure 6c) in the PDMS mold. The geometry of this template is not limited to parallel grooves. Multiple geometries have been explored for 2-D templated assembly, including circles and polygons [25]. The second periodicity, in the iCVD polymer, is attributed to the colloidal template (Figure 6d). This is the smaller length scale that gives rise to the nanostructure. This length scale can be easily tuned by selecting colloidal particles of a different diameter.

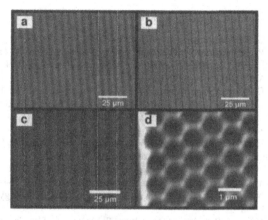

Figure 6. Hierarchical patterns from grafted iCVD polymer were produced with capillary force lithography. A polystyrene template (a) was used to template the colloidal assembly (b). The iCVD polymers (pHEMA shown) are patterned at two length scales. The large length scale corresponds to the polystyrene template (c). The smaller length scale is attributed to the colloidal template (d).

CONCLUSIONS

We present a simple set of techniques for creating large well-ordered arrays of polymeric nanostructures which are covalently bound to the substrate. These structures were templated by a 2-D assembly of a colloidal monolayer and grafted onto the underlying substrate by a bottom-up approach. iCVD allowed us to produce polymeric films which were analogous in composition to the monomer precursor, indicating that identical functionality was realized. Patterns which were created with 1 μm particles generated polymeric dimensions as small as 150 nm in width, and features as tall as 500 nm. These patterns survived long exposure to an aggressive solvent and AFM imaging shows that the patterned films produce a "bowl" structure which matches the hemispherical geometry of the colloidal template. We employed capillary force lithography as a top-down approach for assisting the bottom up assembly. This process is generic and is an inexpensive way to pattern any iCVD polymer to create high surface-area honeycomb patterns that are hierarchically patterned, which is attractive for a variety of applications where spatial registration of functional patterns is desired.

ACKNOWLEDGMENTS

The authors acknowledge the support of the NSF/SRC Engineering Research Center for Environmentally Benign Semiconductor Manufacturing as well as the support of the National Defense Science and Engineering Graduate Fellowship. We would like to thank Sung Gap Im for providing the PDMS molds.

241

REFERENCES

1. Y. Chen, A. Pepin, Electrophoresis 22 (2001) 187-207.
2. H. W. Deckman, J. H. Dunsmuir, Applied Physics Letters 41 (1982) 377-379.
3. K. H. Park, S. Lee, K. H. Koh, R. Lacerda, K. B. K. Teo, W. I. Milne, Journal of Applied Physics 97 (2005).
4. X. D. Wang, C. J. Summers, Z. L. Wang, Nano Letters 4 (2004) 423-426.
5. Z. P. Huang, D. L. Carnahan, J. Rybczynski, M. Giersig, M. Sennett, D. Z. Wang, J. G. Wen, K. Kempa, Z. F. Ren, Applied Physics Letters 82 (2003) 460-462.
6 A. Hatzor-De Picciotto, A. D. Wissner-Gross, G. Lavallee, P. S. Weiss, Journal of Experimental Nanoscience 2 (2007) 3-11.
7 J. C. Hulteen, D. A. Treichel, M. T. Smith, M. L. Duval, T. R. Jensen, R. P. Van Duyne, Journal of Physical Chemistry B 103 (1999) 3854-3863.
8. V. Ng, Y. V. Lee, B. T. Chen, A. O. Adeyeye, Nanotechnology 13 (2002) 554-558.
9. A. Valsesia, P. Colpo, M. M. Silvan, T. Meziani, G. Ceccone, F. Rossi, Nano Letters 4 (2004) 1047-1050.
10. A. Ruiz, A. Valsesia, F. Bretagnol, P. Colpo, F. Rossi, Nanotechnology 18 (2007).
11. D. D. Burkey, K. K. Gleason, Journal of Applied Physics 93 (2003) 5143-5150.
12. C. Rau, W. Kulisch, Thin Solid Films 249 (1994) 28-37.
13. A. Grill, Journal of Applied Physics 93 (2003) 1785-1790.
14. A. Grill, V. Patel, Journal of the Electrochemical Society 153 (2006) F169-F175.
15. T. P. Martin, K. K. S. Lau, K. Chan, Y. Mao, M. Gupta, A. S. O'Shaughnessy, K. K. Gleason, Surface & Coatings Technology 201 (2007) 9400-9405.
16. W. E. Tenhaeff, K. K. Gleason, Advanced Functional Materials 18 (2008) 979-992.
17. T. K. S. Wong, B. Liu, B. Narayanan, V. Ligatchev, R. Kumar, Thin Solid Films 462 (2004) 156-160.
18. K. K. S. Lau, K. K. Gleason, Macromolecules 39 (2006) 3688-3694.
19. J. Y. Cheng, C. A. Ross, H. I. Smith, E. L. Thomas, Advanced Materials 18 (2006) 2505-2521.
20. N. J. Shirtcliffe, M. Stratmann, G. Grundmeier, Surface and Interface Analysis 35 (2003) 799-804.
21. P. Amirfeiz, S. Bengtsson, M. Bergh, E. Zanghellini, L. Borjesson, Journal of the Electrochemical Society 147 (2000) 2693-2698.
22. Y. Xia, B. Gates, Y. Yin, Y. Lu, Advanced Materials 12 (2000) 693-713.
23. K. Chan, K. K. Gleason, Langmuir 21 (2005) 8930-8939.
24. Y. Leterrier, L. Medico, F. Demarco, J. A. E. Manson, U. Betz, M. F. Escola, M. K. Olsson, F. Atamny, Thin Solid Films 460 (2004) 156-166.
25. Y. D. Yin, Y. Lu, B. Gates, Y. N. Xia, Journal of the American Chemical Society 123 (2001) 8718-8729.

Mater. Res. Soc. Symp. Proc. Vol. 1134 © 2009 Materials Research Society   1134-BB08-37

Electroactive Polymer Motors for Aerospace Applications

Keith J. Rebello[1], Margaret A. Darrin[1], Jerry A. Krill[1]
[1]The Johns Hopkins University Applied Physics Laboratory, Laurel, MD, 20723 USA

ABSTRACT

A prototype electroactive polymer (EAP) dielectric motor was built from spring roll elements. By utilizing direct electric control of the artificial muscle elements, the need for complex and bulky gearboxes and transmissions were eliminated. The EAP materials allow for the fabrication of motors much lighter than their conventional counterparts with the capability to operate at high altitudes and oxygen free environments. Performance results and lessons learned for future developments are discussed.

INTRODUCTION

Electric motors and combustion engines are widely used in commercial industries but are not appropriate for all applications. In particular aerospace applications can benefit from lightweight motors which can operate at high altitudes such as in the stratosphere layer of the atmosphere or in the atmosphere of a different planet, such as Mars. For example, there is use for a class of balloon airships which can support a payload at high altitude at a relatively fixed geographic location for extended periods of time. To maintain geographic position in the presence of high altitude winds, where the air has low density, a motor and transmission are useful that can turn a station-keeping propeller that is both large and slowly rotating. Another class of vehicles includes spacecraft and satellites, which need light, efficient technology, for example to extend arms and solar panels or to open or close compartments. Probes delivered by these vehicles—like the Mars rovers or submersible autonomous vehicles for the exploration of extraterrestrial seas also could use lightweight and efficient motors and transmission elements.

The objective of this project was to build an electroactive polymer (EAP) dielectric motor by using spring role elements. These motors have the potential to operate in regions of high magnetic fields in space, but also can be used for terrestrial applications such as in particle accelerators and MRI machines. Spring role elements are capable of generating high strains and forces, at high voltages and low currents. In this configuration a biaxial pre-strained and double side coated EAP film is rolled around an elastic coil. The actuator elongates longitudinally as voltage is applied allowing for large displacements to be generated with high forces. The elastic coil contracts the actuator when voltage is deactivated. Various piston and lever designs have been proposed for making dielectric motors[1]. The motor designs in this paper make use of these properties by arranging the spring rolls in piston configurations.

EXPERIMENTAL DETAILS

EAP spring roll elements and the necessary high voltage amplifiers were acquired from Artificial Muscle, Inc (AMI). A pair of MC01 Muscle Controllers were used for actuation. Three EAP spring roll actuators were built for our application by AMI, based on their R970 spring roll actuator, but without their metal housing. This enabled us to make a smaller metal free motor.

A motor completely free of metal was designed and fabricated shown in figure 1. This design uses a clutch to allow the EAP actuators to independently turn the motor's shaft. This enables multiple actuators to be strung together on a single shaft, allowing the motor to be customized for a specific application. This architecture also allows for the use of an electronic transmission, enabling the actuators to operate in phase to increase torque or out of phase to increase rotational speed. Our prototype motor uses two spring roll actuators to drive the shaft.

Figure 1. Schematic of clutch mechanism (left) fabricated motor (right)

The metal free motor's performance was evaluated using a Mecmesin AFG 50N force meter with a 6 NM smart rotary torque sensor.

DISCUSSION

Despite being commercial parts, the spring roll actuators were made to order early stage devices. As such no data on their performance or lifetimes were available from the company. At the time of purchase AMI was the only commercial supplier of spring roll actuators, however there are now other commercial suppliers. The spring roll actuators were hand built by AMI and thus varied slightly in their composition and dimensions. The company was working on an improved manufacturing technique to improve their consistency and reliability, and shorten their lengthy manufacturing time.

The spring role actuators were characterized using DC measurements, figure 2. They were shown to have an exponential length change to applied voltage–from 1 mm at 2000 V up to 6 mm at 4500 V. The maximum force generated by the actuators was measured as 5.3 N at 4500 V. The maximum drive frequency was measured to be 20 Hz. The frequency, stroke distance, and force are all lower than what was reported in the literature [2]. During testing one of the spring roll actuators failed after only 10 actuations, leaving us with two functional actuators for the motor. It should be noted that Artificial Muscle Inc. no longer sells a spring roll actuator instead favoring their diaphragm based Universal Muscle Actuator (UMA) [3].

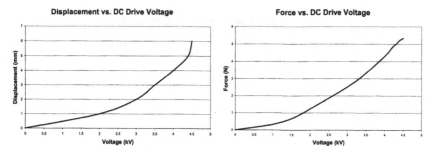

Figure 2. Displacement (left) and Force (right) measurements vs. applied DC drive voltage

Once inserted into the motor the spring rolls, both in phase and out of phase torque measurements were taken, figure 3. Using the two actuators in phase the motor has a stall torque of 146 mNm, which is competitive with conventional electric micromotors. Operating torque was independent of excitation voltage up to 4 kV, upon which higher excitation frequencies resulted in lower torque. This was due to the spring rolls not reaching their full displacement when driven at the higher frequencies. When operating at high voltages the operating frequency may need to be reduced in order to achieve maximum performance from the actuators. The motor achieved a maximum rotational speed of 10 RPM when the two actuators were run out of phase at 8 Hz, figure 4. The motor was most efficient when the two actuators were run out of phase. In the in phase or single actuator configuration there is some backlash in the shaft movement which does not occur in out of phase condition as the shaft is prevented from moving backwards by the out of phase actuator. The output power at 4500 V was 0.27 ft-lbs/sec. This was much less than what was expected and due primarily to the lower than expected performance of the EAP actuators. AMI indicated that the current draw of the EAP spring role actuators is less than 1 mA, which would lead to a power consumption of 4.5 W. As manufacturing processes improve the power consumption should reduce.

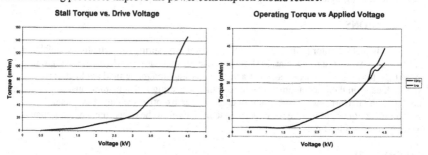

Figure 3. Stall torque (left) and Operating torque (right) measurements vs. applied drive voltage

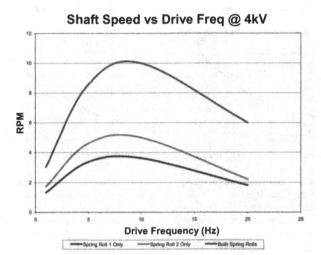

Figure 4. Shaft speed vs. drive frequency at 4 kV. Peak RPM achieved at 8 Hz.

A month after testing was completed the second EAP spring roll catastrophically failed while in storage. It looked as if the spring was exerting too much force and ripped the EAP rolls apart. It is unknown whether too many actuation cycles weakened the EAP sufficiently enough to cause failure. Future efforts should characterize the lifetime of these types of actuators. Diaphragm based EAP technology is simpler, more mature, and is currently capable of operating at higher frequencies than spring rolls, albeit with less force. Possible use of these types of actuator technology should be investigated for future motor applications.

CONCLUSIONS

A dielectric motor built from electro active polymer spring roll actuators was fabricated and tested. EAP spring roll actuators show promise for applications where light weight low speed high torque motors are needed for aerospace or high magnetic field applications. Although early in their development cycle, commercial spring roll actuators must improve their reliability and performance to successfully compete against other technologies. As the technology continues to mature we expect these limitations to be overcome, enabling many exciting future applications.

REFERENCES

1. Jerry A. Krill, US Patent No. 7,071,596 (4 July 2006)
2. S. Ashley, "Artificial Muscles," *Scientific American*, October 2003, pp. 52-59.
3. http://www.artificialmuscle.com

Mater. Res. Soc. Symp. Proc. Vol. 1134 © 2009 Materials Research Society   1134-BB08-42

New Application of AAO Templates: Periodic Patterns of Nanonet Architecture

Zhixun Luo[1], Yuanyuan Liu[1], Longtian Kang[1], Yaobing Wang[1], Hongbing Fu[1], Ying Ma[1], Jiannian Yao[1,*], Boon H Loo [2]

[1]Beijing National Laboratory for Molecular Science (BNLMS), Institute of Chemistry, Chinese Academy of Sciences, Beijing 100190, China. E-mail: jnyao@iccas.ac.cn
[2]Department of Chemistry, Towson University, Towson, MD 21252, USA, E-mail: bloo@towson.edu

ABSTRACT

Self-assembly of organic molecules from solution is one of the simplest methods to generate ordered nanostructures with potentially new properties. Template synthesis has been widely used as a controllable approach to achieve desirable nano-structured materials. Of the many different types of templates, the anodic alumina oxide (AAO) offers obvious advantages in the making of one-dimensional nanostructured materials and arrays, though the AAO has been used as a template for the syntheses of one dimensional nanomaterials and nanostructures, there have been few reports on the application of the AAO as a substrate for surface morphology control in macroscopic scale. In this communication, we report the synthesis of large-area (ca. 12 cm^2) nano-net architecture of TClPP (5, 10, 15, 20-tetrakis (p-chlorophenyl) porphyrin, $C_{44}H_{26}Cl_4N_4$) using the AAO template as a substrate, utilizing pressure difference method. We demonstrated the pin-hole migration mechanism for the nano-net assembly based on a controllable dynamic balance. The nano-net architectures may have potential applications in catalyst and interface investigations, and may extend new applications of AAO templates in two-dimensional organic nanonet structures.

Key words: Controllable assembly; Nano-net, Pressure difference, AAO template, Porphyrin

INTRODUCTION

Generation of molecular architectures that exhibit a high degree of order across multiple length scales is significant for application in various fields including (opto) electronics, magnetism, catalysis, and medicine[1-7]. Alkylated polycyclic discotic molecules such as porphyrins are frequently chosen as starting building blocks to exploit their ability to stack and form architectures and liquid-crystalline phases. Self-assembly of the organic molecules from solution is one of the simplest methods to develop complex, nanostructured materials with innovative properties. However, the controllable process is still limited by available fabrication methods.

Template synthesis has been widely used as a controllable approach to achieve desirable nano-structured materials. Of the many different types of templates, the anodic alumina oxide (AAO) [8-14] offers obvious advantages in the making of one-dimensional nanostructured materials and arrays; the AAO templates provide hexagonally packed, uniform pore arrays with pore diameter that can be varied up to 200 nm. Amongst other applications, the AAO has been used as a template for the syntheses of nanotubes for biomedicine and biotechnologogy[10], $Bi_{1-x}Sb_x$ nanowires as thermoelectric wires[11], SBA-15 nanorod arrays for protein separation and catalysis[12], and lipid nanotube arrays as a model of cellular membranes[14]. Despite these progresses, there have been few reports on the application of the AAO as a substrate for surface morphology control in macroscopic scale.

In this paper, we report the synthesis of large-area (ca. 12 cm^2) nano-net architecture of TClPP (5, 10, 15, 20-tetrakis (p-chlorophenyl) porphyrin, $C_{44}H_{26}Cl_4N_4$) using the AAO template as a substrate. Scheme 1 gives the 3-D structure of the TClPP molecule and its stacking.

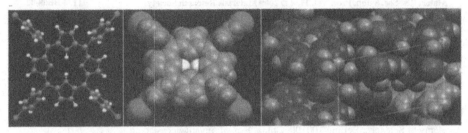

Scheme 1   The 3-D structure of the TClPP molecule and its stacking.

EXPERIMENTAL

AAO templates were prepared according to the method proposed by Masuda et al. Aluminum foils (99.999% purity) were bought from Chinese General Research Institute for Nonferrous Metals. Before anodizing, raw cooling rolling aluminum foils were first degreed in acetone and then annealed in furnace at different temperatures for 2.5 hours.

Annealed aluminum foils were first oxygenated at a constant voltage of 40 V in 0.4M oxalic acid at the temperature below 10 °C for 2-3 hours. After removing the aluminum oxide layer formed at the above step in a mixture of phosphoric acid (6wt %) and chromic acid (1.5wt %), the aluminum foils were subjected to an anodic-oxidation treatment again at the same conditions. Copper chloride solution is used to peel off the aluminum base and phosphoric acid is used for enlarging the pores.

The dimension of AAO pores can be controlled between 20 and 120nm on different preparation conditions according to the method used. The following SEM pictures (Fig.1a and Fig.1b) give two AAO templates with 50nm and 100nm pores prepared based on this method; AAO templates with 200nm pores are purchased from Whatman (Fig.1c).

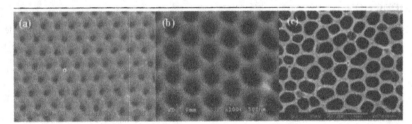

Fig. 1.   SEM images of the AAO templates with pores of (a) 50nm, (b) 100nm and (c) 200nm.

The TClPP nano-nets were fabricated as follows: A 2-cm diameter AAO disk of thickness ca. 15 μm was laid on a Buchner funnel fitted with a fritted disc. The funnel was then placed on a filter flask connected to a vacuum pump, which was used to maintained a pressure differential across the AAO disk. 1 mL of 0.13 M TClPP in CH_2Cl_2 solution was added dropwise onto the AAO template. TClPP nano-nets of different meshes in accordance with the AAO pore sizes were thus grown on the back side of the AAO template, i.e., the side opposite to that of TClPP deposition. This process was repeated for several times to obtain nanoparticles of larger size. Experimental parameters, such as the pressure differential, concentration of the TClPP solution and the number of times of solution deposition, affected the type and the quality of the nano-structures formed, and they will be discussed in the following.

The morphology and size of the sample were examined by field emission scanning electron microscopy (FESEM, Hitachi S-4300). The optical properties of the sample were investigated by steady-state fluorescence spectrometer (Hitachi F-4500) and by UV-VIS absorption spectrometer

(Lambda 35 1.27 spectrometer). The Raman spectra were collected on a *Renishaw* 2000 Raman spectrometer, with an excitation wave length at 514.5nm.

RESULTS AND DISCUSSION

Figure 2 shows the as-prepared periodic network patterns of TClPP from an AAO template. As one can see, small nanoparticles assembled to form a nanonet structure. The results indicate that an AAO template with small pores assisted the formation of a pattern structure on the back of the AAO template.

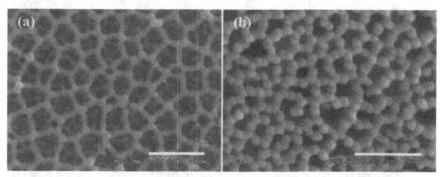

Fig. 2. SEM images of TClPP nanonets cultivated on the AAO templates with pressure difference method (the dimension bar is 500nm).

By directly dropping TClPP solutions on a flat silicon surface, one can only obtain a un-ordered film with some non-uniform rings. Actually evaporation typically proceeds through dewetting of solution, leading to the formation of holes in the liquid layer that eventually enlarge and coalesce until the solvent is completely evaporated. This process is typically called a pinhole mechanism. The morphology based on the AAO template is dependent of the interfacial interaction, referred to as the molecule-substrate effect[1]. It is remarkable that the nanonet architecture is uniformly self-assembled with interlocking nanoparticles. The mesh knots are formed by a single larger nanoparticle or even several smaller nanoparticles. It was supposed that the TClPP molecules go through the AAO pores and form bubbles, dewetting with bubbles confluence or broken. Synchronously, the molecules will self-assemble, interacting with one another through weak van der Waals or electrostatic interactions. The non-covalent interaction can also be directional, with solvent evaporation molecules can arrange on the pore edge to form periodic patterning such as the nanonet architectures. There is even a competition between the self-assembly and dewetting[15]. It is notable that the pressure difference is the essential factor to obtain the nanonet architecture. Without pressure difference (or just by means of a soaking method based on the same AAO template) we could not obtain the nanostructures.

Figure 3 gives more elaborate SEM images of TClPP nano-nets cultivated on AAO template by means of pressure difference methods. It is clearly that the as-prepared nano-net can be well shape selective and shape controlled. Essentially, the final nanonet morphology actually results from the interplay of intramolecular, intermolecular, and interfacial interaction, and the surface tension forces owing to molecule-solvent. To obtain a controllable morphology, a subtle balance of all the interactions involved must be achieved[1]. We have found that smaller molecules were relatively more difficult to form the uniform nanonet architectures. This suggests that the intramolecular and intermolecular interactions are the first necessary conditions. Equilibrium between dewetting and intermolecular interaction is the major driving force of the controllable assembly.

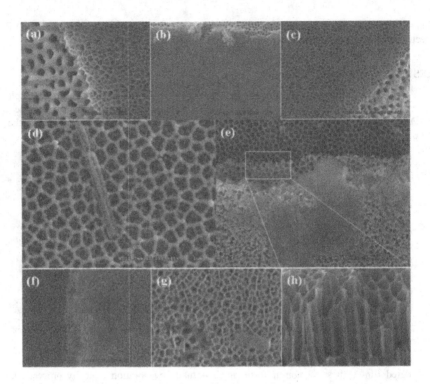

Fig. 3. SEM images of TClPP nano-nets cultivated on AAO template by means of pressure difference methods. (a~c refer to the characters at the adding edge; d refers to 100nm AAO, 0.04MPa; e~h aim at the situation at some special area)

The absorption experiments performed on a Lambda 351.27 spectrometer revealed that the TClPP nano-net architecture exhibits different absorption chracteristics from that of TClPP in solution, as shown in Fig. 4. The TClPP in solution shows a strong monomer absorption at 418 nm, assigned to the Soret band, in addition to four barely visible peaks at 515, 549 590, 645 nm, assigned to the Q band (Fig. 4a)[15, 16]. The TClPP nano-net, in contrast, shows a total of six peaks of strong intensity in the region (Fig. 4b). Except for a small wavelength shift from the corresponding peak in solution TClPP, the 415-nm band can be assigned to the Soret band. Likewise, the 517-, 552-, 593-, 649-nm peaks are assigned to the Q band. The enhancement in the intensity of the Q band is attributed to a lowering of molecular symmetry of TClPP in the nano-net architecture because of molecular aggregation. The most interesting feature in the absorption spectrum of the TClPP nano-net architecture, however, is the appearance of the 452-nm peak. The new 452-nm peak can be attributed to J-aggregation of the TClPP molecules in a head-to-tail arrangement [18]. Hence, the absorption results suggest the occurring of J-aggregation in the TClPP nano-nets.

Figure 5 gives the Raman spectra of TClPP from a nanonet adhered on AAO template (Fig.5a) and from a floating AAO nanonet without AAO support (Fig.5b), respectively. It is needed to mention here that the bulk TClPP (purple powders) shows very poor Raman signal due to its very strong adsorption. Herein we achieved the Raman spectra from the AAO-based nanonet architectures

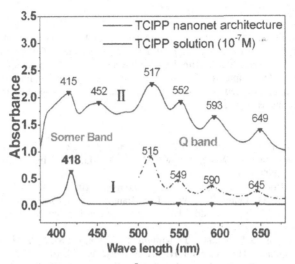

Fig.4. (a) The absorption spectrum of 0.5 x 10^{-7} M TCIPP in CH_2Cl_2 solution, showing a strong Sorent band at 418 nm, and four very weak peaks at 515, 549, 590, 645 nm (referred to as the Q band and they are manified immediately above); (b) The absorption spectrum of the TCIPP nano-net architecture resembles that of solution TCIPP (spectrum a), except the presence of an additional peak at 452 nm.

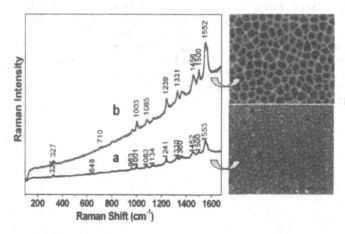

Fig. 5. Raman spectra of TCIPP nano-net architectures: (a) from a nanonet adhered on AAO, (b) from a floating AAO nanonet

Although very regular patterns have been obtained in some super-molecule system[15], the interplay of the different forces is truly complex and thus the formation of reproducible morphologies requires a high level of control over many deposition parameters such as the drop size, drop-edge pinning, temperature, humidity, nature of the surface, and so on. However, by employing the AAO template, it takes advantages of the AAO periodic structure to give these deposition parameters to the simple control of characters of AAO template and the pressures.

251

Conclusion

A large network (ca. 12 cm^2) of nano-net architecture of TClPP has been fabricated on the AAO templates by applying a pressure differential across the deposited TClPP films on the templates. Stable nano-nets can be reproduced with the same mesh patterns of the hexagonally packed, uniform pore arrays of the AAO templates. While the extension of the present findings to other molecules or systems remains to be explored, the unique structural features of the nano-nets may find applications in molecular filtering as well as in catalysis.

References

(1) Palermo,V.; Samori, P. *Angew. Chem., Int. Ed.* **2007**, *46*, 4428

(2) Longtian Kang, Zhechen Wang, Zongwei Cao, Ying Ma, Hongbing Fu, and Jiannian Yao□*J. Am.Chem. Soc.* **2007**, 129, 7305

(3) Bertorelle, F.; Rodrigues, F.; Fery-Forgues, S. *Langmuir* **2006**, *22*, 8523

(4) Hu, J. S.; Guo, Y. G.; Liang, H. P.; Wan, L. J.; Jiang, L. *J. Am. Chem. Soc.* **2005**, *127*, 17090

(5) Balakrishnan, K.; Datar, A.; Naddo, T.; Huang, J.; Oitker, R.; Yen, M.; Zhao, J.; Zang, L. *J. Am. Chem. Soc.* **2006**, *128*, 7390

(6) Kaneko, Y.; Shimada, S.; Fukuda, T.; Kimura, T.; Yokoi, H.; Matsuda, H.; Onodera, H.; Kasai, H.; Okada, S.; Oikawa, H.; Nakanishi, H. *Adv. Mater.* **2005**, *17*, 160

(7) Kim, S.; Zheng, Q. D.; He, G. S.; Bharali, D. J.; Pudavar, H. E.; Baev,A.; Prasad, P. N. *Adv. Funct. Mater.* **2006**, *16*, 2317

(8) Masuda, H.; Fukuda, K. *Science* **1995**, 268, 1466

(9) Cui, Y.; Lieber,C.M. *Science* **2001**, 291, 851

(10) Gasparac, R.; Kohli, P.; Mota, M.O.; Trofin, L.; Martin, C.R. *Nano letters*, **2004**, 4, 513

(11) Prieto, A. L.; Gonzalez, M.M.; Keyani, J.; Gronsky, R.; Sands T.; Stacy A. M. *J. Am.Chem. Soc.***2003**,125, 2388

(12) Lu, Q.Y.; Gao F.; Komarneni S.; Mallouk, T. E. *J. Am.Chem. Soc.* **2004**, 126, 8650

(13) Chia, S.; Urano, J.; Tamanoi F.; Dunn, B.; Zink, J.I. *J. Am. Chem. Soc.* **2000**, 122, 6488

(14) Smirnov A.I.; Poluektov, O.G. *J. Am. Chem. Soc.* **2003**, 125, 8434

(15) Hameren, R.V.; Schon,P.; Buul,A.M.; Hoogboom,J.S.; Lazarenko,V.; Gerritsen,J.W.; Engelkamp,H.; Christianen, P.C.M.; Heus,H.A.; Maan,J.C.; Rasing,T.; Speller,S; Rowan, A.E. Elemans, J.A.A.W.; Nolte,R.J.M. *Science*, **2006**, 314, 1433

(16) Luo, Z.; Liu, Y.; Kang, L.; Wang, Y.; Fu, H.; Ma, Y.; Yao, J.; Loo, B. H. *Angew. Chem. Int. Ed*, 2008, **47** 8905 - 8908.

Mater. Res. Soc. Symp. Proc. Vol. 1134 © 2009 Materials Research Society   1134-BB08-46

Synthesis of Thermoresponsive Copolymers Composed of Poly(ethylene glycol) and Poly(N-isopropyl acrylamide) for Cell Encapsulation

Tatiya Trongsatitkul and Bridgette M. Budhlall

NSF NSEC Center for High-Rate Nanomanufacturing, Department of Plastics Engineering, University of Massachusetts-Lowell, Lowell, MA 01854, U.S.A.

ABSTRACT

Thermoresponsive copolymers of poly(N- isopropyl acrylamide) (PNIPAm) and poly(acrylamide) microgels copolymerized with poly(ethylene glycol)(PEG) chains were synthesized by free-radical photopolymerization. Poly(ethylene glycol) methyl ether methacrylate (PEGMA) macromonomers with varying number-average molecular weights were used (M_n = 300 and 1,000 g/mol). A simple microarray technique coupled with a laser scanning confocal microscope (LSCM) was used to visualize the effect of temperature on the volume phase transition temperatures of the microgels. In general, increasing the concentration of PEGMA in the PNIPAm-co-Am-co-PEGMA copolymers resulted in a broader and higher volume phase transition temperature (PVTT) compared to the PNIPAm microgels. We demonstrated that the PEGMA molecular weight and concentration influenced whether it was incorporated as a grafted copolymer or random copolymer in the PNIPAm microgel. The evidence for this is the shift in PVTT as determined by temperature response and differential scanning calorimetry (DSC) measurements. This behavior suggests that incorporation of PEGMA in the copolymer depends on its hydrophilicity or water-solubility which in turn influenced the degree at which the copolymer chains collapsed from a coil-to-globule (volume phase transition) with increasing temperature.

INTRODUCTION

Poly(N- isopropyl acrylamide) (PNIPAm) is a well known temperature responsive polymer. PNIPAm demonstrates a volume phase transition temperature (VPTT) or lower critical solution temperature (LCST) in aqueous solution at about 32°C [1]. At temperatures above LCST, the PNIPAm chains collapse due to dissociation of water as PNIPAm becomes more hydrophobic. PNIPAm microgels with a characteristic LCST close to biological temperatures is attractive for various biomedical applications including controlled drug delivery release, tissue regeneration and cell encapsulation.

Cell encapsulation is one of the most promising potential biomedical applications of PNIPAm microgels. In cell encapsulation, transported cells are protected from immune rejection by an artificial, semi-permeable membrane, potentially allowing transportation without the need for immuno-suppression [2]. To improve the biocompatibility of PNIPAm microgels, surface modification is needed to render stealth properties to the microgel by coating or grafting with biocompatible polymers, such as poly(ethylene glycol)(PEG) or poly(ethylene oxide)(PEO). The stealth character of PEG is based on its steric repulsion properties. In aqueous solution, the oxygen on PEG segments form hydrogen bonds with water, providing the PEG segments with a protective hydration shell that minimizes the non specific interactions of the modified microgels within the biological environment [2, 3].

Incorporation of PEG with PNIPAm microgels achieves synergist benefits of biocompatibility from PEG and thermal responsive characteristics from PNIPAm, respectively.

However, PEG incorporation may alter the LCST or VPTT of PNIPAm. LCST is principally dependent on the hydrogen bonding capability of the constituent monomer units. Thus, incorporating PEG can be expected to change the hydrophilic/hydrophobic characteristic of PNIPAm leading to a shift in the LCST or VPTT with increasing temperature.

The goal of this research is to understand the effect of grafting PEG macromonomer to PNIPAm on the volume phase transition temperature of the grafted copolymer microgel. These PNIPAm-co-PEGMA copolymers will then be explored as thermoresponsive microgels for encapsulating mammalian cells for controlled delivery and release applications.

EXPERIMENT

Materials

All of the chemicals used in these experiments were obtained from Sigma Aldrich and used as received, unless otherwise noted. N-isopropylacrylamide (NIPAm), acrylamide (Am), poly(ethylene glycol) methyl ether methacrylate (PEGMA) (M_n = 300 and 1,000), N,N'-methylenebisacrylamide (BIS), 2,2-dimethoxy-2-phenylacetophenone (oil-soluble photoinitiator), anthraquinone-2-sulfonic acid, sodium salt monohydrate (water-soluble photoinitiator), fluorescein isothiocyanate (FITC) dye, mineral oil, deionized water (Millipore Elix® 3).

Synthesis of PNIPAm-co-PEG

In this research, the effect of incorporating PEG macromonomer with PNIPAm-co-Am on the volume phase transition temperature (PVTT) of the microgel was investigated. The number average molecular weight (M_n) and concentration of PEGMA were varied. N-isopropylacrylamide (NIPAm) was copolymerized with acrylamide (AM) in a 95:1 molar ratio to increase the LCST of the resulting PNIPAm-co-Am copolymer temperatures closer to body temperature [4]. The effect of varying the molecular weight of PEGMA macromonomer at three different concentrations (10, 20, and 30 wt%) was also investigated. A crosslinker, N,N'-methylenebisacrylamide (BIS) in a molar ratio of 1:750 (crosslinker: monomers) was used to increase the molecular weight of the copolymer in addition to render the copolymer water-insoluble Oil-soluble and water-soluble photoinitiators (1:1 molar ratio) at 3000 ppm (based on monomer) were used to initiate the polymerization reaction. FITC dye was used to label the monomer droplet under epifluorescence mode of LSCM.

A simple microarray technique was used that enabled real-time observation of the polymerization and the thermal response of microgel under the optical microscope. As such, only a small quantity of chemicals is required for an experiment using this method. This is a major advantage to using this technique for exploratory research. A schematic of the microarray technique used to synthesize the PEG –g-PNIPAm microgels is shown in Figure 1. First, mineral oil was placed in a concave well on a poly(dimethyl siloxane) (PDMS) coated microslide. Second, a small drop (0.8 µl/drop) of monomer solution, prepared as described above, was placed onto the mineral oil. The monomer solution drops initially sink into the oil and serve as self-contained micro-reactors. Photopolymerization was conducted using UV irradiation. Observations under LSCM were then made every 30 mins. The microgels were completely polymerized after 90 mins of UV irradiation.

The temperature response of the microgels was then characterized by observations under the LSCM. A hot-stage (Instec WS60) was used to vary and control the temperature of the microslides containing the microgels. The temperature was varied from 25 to 55°C at 5°C increments. The microgel samples were allowed to reach temperature equilibrium for 10 mins before microscope images were acquired. These images were then analyzed to determine the microgel diameters for comparison.

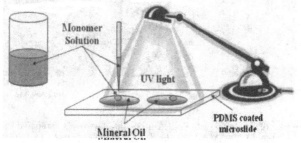

Figure 1. Schematic of microarray technique used for real-time observation of the polymerization

DISCUSSION

The VPTT of PNIPAm can be varied by copolymerization. In the present study, acrylamide was incorporated to achieve a PNIPAM-*co*-Am microgel with VPTT in the range of 38-41°C which is slightly above physiological body temperature of 37°C [5]. By incorporating another hydrophilic monomer-PEGMA, it is expected that the VPTT will be further increased. The extent of increase in VPTT can be adjusted by varying the concentration of the hydrophilic monomer. A summary of the experiments performed with PNIPAm-*co*-Am microgels synthesized using PEGMA with two different molecular weights and various concentrations is shown in Table I.

Table I. Experimental Parameters and Observations for PNIPAm-*co*-Am Microgels Copolymerized with PEGMA

Property	Control PNIPAm-*co*-Am	Low molecular weight-PEGMA (M_n = 300 g/mol)			High molecular weight-PEGMA (M_n = 1,000 g/mol)		
		10 wt%	20 wt%	30 wt%	10 wt%	20 wt%	30 wt%
VPTT	~35 °C	~40°C	~45°C	~50°C	~40°C	~35°C	~32°C

The effect of PEGMA number average molecular weight (M_n) on the VPTT of the grafted PNIPAm microgels was studied. A series of microgels were prepared without PEGMA and with 300 and 1,000 g/mol PEGMA respectively. They were all dispersed in a continuous phase of mineral oil as illustrated in Figure 1. The effect of PEGMA concentration on the volume phase transition of the PEGMA-*co*-Am microgel after exposure to temperatures ranging from 25-55°C was also studied. The behavior of these microgels suspended in mineral oil is illustrated for only (3) temperatures in Figure 2.

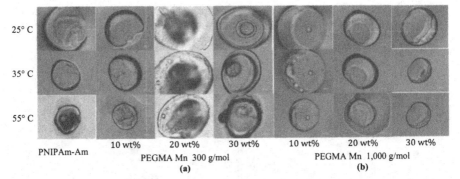

| | 10 wt% | 20 wt% | 30 wt% | 10 wt% | 20 wt% | 30 wt% |

25° C

35° C

55° C

PNIPAm-Am   PEGMA Mn 300 g/mol   PEGMA Mn 1,000 g/mol

                 (a)   (b)

Figure 2. Optical micrographs of PNIPAm-*co*-Am microgels prepared with increasing concentration of (a) low M_n PEGMA (300 g/mol) and (b) high M_n PEGMA (1,000 g/mol).

    In general, it was observed that the dimensions of all the microgels decreased with increasing temperature. The normalized difference in diameter is plotted as a function of temperature for the entire temperature range measured (25-55°C) for PNIPAm microgels copolymerized without and with PEGMA macromonomers with low M_n (Figure 3) and high M_n (Figure 4), respectively.

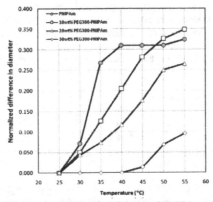

Figure 3. The comparison of temperature-induced shrinkage of PNIPAm microgels with 0 wt% (○), 10 wt% (□), 20 wt% (Δ), and 30 wt% (◊) PEGMA (M_n = 300g/mol).

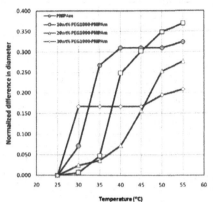

Figure 4. Comparison of temperature-induced shrinkage of PNIPAm microgels with 0 wt% (○), 10 wt% (□), 20 wt% (Δ), and 30 wt% (◊) PEGMA (M_n=1,000g/mol).

The VPTT of the microgels prepared with low M_n PEGMA increased and broadened with increasing concentration. In contrast, the VPTT of the microgels prepared with high M_n PEGMA, decreased and sharpened with increasing concentration in the range of 10-30 wt%. This behavior can be explained by the decrease in tendency for the hydrophobic associations of isopropyl groups of PNIPAm segments when hydrophilic comonomer units are incorporated. This in turn led to a broader VPTT that occurred at a higher temperature [3]. Similar trends were also found when other hydrophobic comonomers were incorporated into PNIPAm [1, 3, 6, 7]. When the highest concentration (30 wt%) PEGMA macromonomer at the highest molecular weight (Mn = 1,000 g/mol) was used; the VPTT persisted at 32 °C, a temperature closer to the characteristic LCST of pure PNIPAm. A plausible hypothesis for these results is that the M_n influences the miscibility of the two comonomers. As the M_n of PEGMA increased from 300 g/mol to 1,000 g/mol, the PEGMA became more hydrophobic and therefore less soluble. At higher concentrations this immiscibility enhanced the phase separation of PNIPAm. Therefore, the VPTT shown is closest to the LCST of pure PNIPAm due to its greater insolubility at the highest concentration and phase separation. This hypothesis is further confirmed by DSC as shown in Figures 5 (a) and (b).

The DSC thermograph of PNIPAm-co-Am (Figure 5 (a)) reveals an endothermic peak at 33.7 °C indicating the VPTT of PNIPAm-co-Am. As expected, incorporating PEGMA (M_n=300 g/mol) resulted in increasing of VPTT to ~45 °C with a broader transition. This result is in agreement with the observations made using LSCM. The single peak found in the DSC themographs of PNIPAm-co-Am copolymerized with PEG (M_n=300 g/mol) suggests that the resulting product is "random" copolymer. On the other hand, thermographs of PNIPAm-co-Am copolymerized with PEGMA (M_n 1,000 g/mol) (Figure 5 (b)) shows two endothermic peaks, one at ~32 °C and the other at ~45 °C. These (2) peaks are well defined in the thermograph of PNIPAm-Am with the highest concentration of PEGMA (1,000 g/mol), 30 wt%. This indicates that the obtained copolymers using higher Mn PEGMA are a mix of "*random*" copolymer and "*block*" copolymer. The DSC thermographs confirm the effect of molecular weight on phase separation of the two comonomers on the copolymerization mechanism as hypothesized.

257

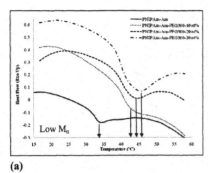 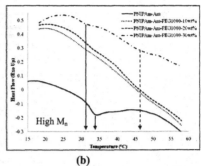

(a)   **(b)**

Figure 5. DSC thermographs of PNIPAm-*co*-PEG micogels (a) PEG (M_n=300g/mol) and (b) PEG (M_n=1,000 g/mol)

CONCLUSIONS

PNIPAm-*co*-PEGMA microgels were synthesized using free radical photo-polymerization via microarray technique. Incorporation of PEGMA macromonomer was found to broaden and shift the volume phase transition temperature toward a higher temperature as compared to the characteristic LCST of PNIPAm. It is believed that the PEGMA increased the hydrophilicity of the PNIPAm-*g*-PEGMA microgels and hindered the association of the isopropyl groups of PNIPAm segments, resulting in a higher LCST for the low M_n PEGMA. The poor solubility of the comonomers is believed to play an important role in the phase transition temperature of the microgels as illustrated by the highest concentration of PEGMA macromonomer (M_n = 1,000 g/mol). These microgels exhibit a similar phase transition temperature of about 32°C, close to that of the control (pure) PNIPAm microgels.

REFERENCES

[1] J. Virtanen, C. Baron, and H. Tenhu, "Grafting of Poly(N-isopropylacrylamide) with Poly(ethylene oxide) under Various Reaction Conditions," *Macromolecules,* vol. 33, pp. 336-341, 2000.
[2] G. Orive, R. M. Hernandez, A. R. Gascon, R. Calafiore, T. M. S. Chang, P. D. Vos, G. Hortelano, D. Hunkeler, I. Lacik, A. M. J. Shapiro, and J. L. Pedraz, "Cell encapsulation: Promise and progress," *Nat Med,* vol. 9, pp. 104-107, 2003.
[3] D. Gan and L. A. Lyon, "Synthesis and Protein Adsorption Resistance of PEG-Modified Poly(N-isopropylacrylamide) Core/Shell Microgels," *Macromolecules,* vol. 35, pp. 9634-9639, 2002.
[4] G. A. M. S. R. Sershen, M. Ng, N. J. Halas, D. J. Beebe, J. L. West,, "Independent Optical Control of Microfluidic Valves Formed from Optomechanically Responsive Nanocomposite Hydrogels," *Advanced Materials,* vol. 17, pp. 1366-1368, 2005.
[5] R. Weissleder, "A clearer vision for in vivo imaging," *Nat Biotech,* vol. 19, pp. 316-317, 2001.
[6] C. D. Jones and L. A. Lyon, "Synthesis and Characterization of Multiresponsive Core&Shell Microgels," *Macromolecules,* vol. 33, pp. 8301-8306, 2000.
[7] N. N. Janevieve A. Jones, Kendra Flagler, Christina D. Pagnucco, Steve Carew, Charles Cheong, Xiang Z. Kong, Nicholas A. D. Burke, Harald D. H. Stöver,, "Thermoresponsive copolymers of methacrylic acid and poly(ethylene glycol) methyl ether methacrylate," *Journal of Polymer Science Part A: Polymer Chemistry,* vol. 43, pp. 6095-6104, 2005.

E-NSF and New Materials

Mater. Res. Soc. Symp. Proc. Vol. 1134 © 2009 Materials Research Society 1134-BB09-04

Liquid Crystal Nano-Particles, LCNANOP – A SONS II Collaborative Research Project

John W. Goodby[1], Martin Bates[1], Isabel M. Saez[1], Ewa Gorecka[2], Heinz-S. Kitzerow[3], Daniel Guillon[4], B. Donnio[4], Jose-Luis Serrano[5], Robert Deschenaux[6]

[1]Department of Chemistry, University of York, Heslington, York, YO10 5DD, UK
[2]Department of Chemistry, University of Warsaw, Zwirki I Wigury 101, 02-089 Warsaw 00927, Poland
[3]Department of Chemistry, Faculty of Science, University of Paderborn, Warburger Str. 100, 33098 Paderborn, Germany
[4]Institut de Physique et Chemie de Materiaux, Université Louis Pasteur, 23 rue de Loess, BP 43, 67084 Strasbourg, France
[5]Departamento de Quimica Organica, Facultad de Ciencias, Instituto de Nanociencia de Aragón, Universidad de Zaragoza, 50009, Spain
[6]Université de Neuchâtel, Avenue de Bellevaux 51, Case postale 158, 2009 Neuchâtel, Switzerland

ABSTRACT

LC-NANOP is an ESF EUROCORES SONS Collaborative Research Project that is addressing an innovative approach to self-organized nanostructures by combination of a variety of organic, inorganic and metal scaffolds with the unique self-organization properties of liquid crystals to obtain *liquid crystal nano-particles*. LC-NANOP is concerned with the synthesis, analysis, characterization, modeling and physico-chemical properties of super- and supra-molecular systems which are formed from a nano-particle as a central scaffold, surrounded by a layer of liquid crystal. The self-organization properties of the liquid crystal coating is the driving force leading to the self-assembly of the nano-particles into secondary or tertiary hierarchical structures, with emphasis on the systematic variation of nano-particle size, chirality, shape and functionality. This bottom-up approach to nano-structuring is very powerful as it combines the extraordinary variety of morphologies that liquid crystals present with the combination of functional entities, relevant for chemical, biological, optoelectronic, and photonic tasks, etc, to create ordered nano-structures that can be controlled by external stimuli.

INTRODUCTION

The development of liquid-crystalline nano-particles began in earnest only recently with the synthesis and characterization of mesomorphic dendrimers. Great interest has been shown recently with respect to the various molecular motifs that will support the creation of dendritic and hyperbranched liquid crystal polymers [1-3]. Although an exhaustive revision of this area is too extensive for this article, several types of molecular architecture can be identified as follows; a) hyperbranched polymers and LC dendrimers in which the mesogenic groups form part of each branching unit; b) dendrimers without mesogenic groups that form LC phases, where the formation of mesophases is due to self-assembly processes of the constituent dendrons, aided by microphase segregation, and c) LC dendrimers formed by attachment of LC moieties to the periphery of a suitable core dendrimer/scaffold, where the mesogenic units are located on the surface of the dendritic or scaffold core. It is this latter group of super- and supra-molecular materials that will be the primary focus of this report.

In this area of super- and supra-molecular liquid crystals, many different expressions have been used to describe some of the same structures, however, for the purposes of this article supermolecular describes a giant molecule made up of covalently bound smaller identifiable components, see figure 1(a), and a supramolecular material means a system made up of multiple components that are not covalently bound together, see figure 1(b).

(a)   (b)

Figure 1. (a) Supermolecular entities are covalently bound, whereas (b) supra-molecular systems are bound together by ionic or H-bonding interactions.

Interest in dendritic liquid crystal systems has now **importantly shifted** towards the synthesis of **"functional" materials** because of their rich supramolecular chemistry, self-organizing and self-assembling properties; and the possibility of accessing the very properties of precise control over molecular engineering, enjoyed by low molar mass liquid crystals; for controlling and fine-tuning physical properties which defines the self-organizing process that leads to mesophase formation; control over long-range positional ordering; and subsequent practical applications [4,5].

The primary challenges of the LC-NANOP program has been the rational design, synthesis (pure and/or with up-scaling) and characterization of super- and supra-molecular materials with in-built functionalities, which self-organize and/or self-assemble in order to yield novel materials or states of matter of practical importance.

Liquid-crystalline nano-particles have been designed, with the aid of simulations, in the form of a nano-particle (eg, silsesquioxanes, fullerenes, carbosilanes, gold, silver, titania, viruses and spores etc) as the central scaffold, and where the scaffold was potentially multilayered. Surrounding the scaffold a "liquid-crystalline coat" was incorporated, that was derived from spherical, disc- or rod-like mesogenic units, see figure 2. The external coat was designed to consist of one or more mesogenic layers, which in turn were allowed to accommodate further functional units. The mesogenic coat, however, has the specific purpose of providing the self-organization, and ultimately the self-assembling vehicle for the core nano-particles. Although shown as spherical, the scaffolds do not necessary have to be so, eg gold nanoparticles exhibit various polyhedral forms. Furthermore, they can be designed to have holes and cavities within their structures, thereby allowing formation of ion channels and binding sites. In addition, liquid crystalline nano-particles can be defined as either supermolecular or supramolecular systems, see figure 2.

As noted a **supermolecule** (figure 2 top-right) is a giant molecular entity made up of covalently bonded, molecular units and thus it is similar in constitution to that of a tertiary structure of a *protein*. In contrast, a **supramolecular** system (figure 2 bottom-right) is a self-assembled, non-covalently bonded entity, in which molecular units are assembled through non-covalent forces to create a complex structure, similar in constitution to that of a quaternary structure of a *protein*. Consequently, a supermolecule has a clearly defined chemical structure, whereas for a supramolecular system it is possible to have variations in the constitution depending on how many molecular units are required to create the self-assembling, and hence, self-organizing, complex system.

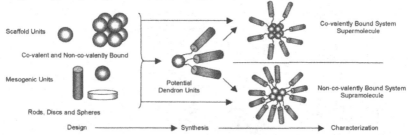

Figure 2: Super- and supra-molecular nano-particles can be prepared from a variety of scaffolds and mesogenic groups, to the right the mesogenic units are shown as rods.

Figure 3 shows more specific templates for supermolecules. The central scaffold is substituted at its surface with mesogenic units as depicted by ellipsoids in (3a). When the mesogenic units are identical a dendritic structure results which can be subject to polydispersity. Where the mesogens are different, as shown by the various structures in 3(b), a discrete polypedal object having many feet is obtained. Thus there is a clear distinction between these two types of supermolecular system. The third example 3(c) is that of a supermolecular functional material where the ellipsoids are mesogenic groups that are used in the self-organizing process, whereas the two groups **A** and **B** are units that possess a particular function.

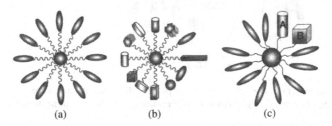

Figure 3: Dendritic (a), discrete molecular (b), and functional supermolecular materials.

The variety of supermolecular materials that can be designed and created produces a variable zoo of structures. The types of self-organizing mesophases formed are dependent on the overall molecular topology of these giant systems. The possibilities for design also create a clash between conventional liquid crystal structure based on low molar mass materials, and those based on polymeric systems. For example two extremes for design are possible, one where the mesophase formation is dependent on shape and microphase segregation, and another where low molar mass mesogenic groups are incorporated into the structure to drive mesophase formation. In between a wide variety of possibilities and combinations become possible. Some examples of these materials design are shown in figure 4.

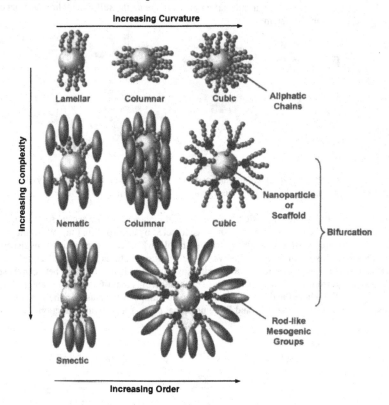

Figure 4. The effect of peripheral units on the packing distributions, curvature and complexity in the self-assembly and self-organization of supermolecular liquid crystalline nano-particles.

EXPERIMENT AND DISCUSSION

Shape Dependency

Although the schematic drawings of liquid-crystalline dendrimers and nano-particles tend to show the materials as being essentially spherical, in reality mesogenic density at the surface of the super- or supra-molecular system, coupled with the liquid crystal environment, can have powerful effects on molecular topology, for example spherical systems can be deformed into rod-like materials because of the mesogenic field. The increased development of theoretical modeling of shape-dependent condensed phases is a necessity for a more complete understanding of such material systems, and critical to the ultimate design processes.

For example, the hexa-adducts of [60]fullerene can give a spherical distribution of mesogenic substituents about the central scaffold, however the fullero-dendrimers exhibit lamellar mesophase behavior. Compound 1 exhibits a relatively high glass transition temperature, 80 °C, corresponding to the formation of a smectic A phase. No other liquid crystal phase is observed and the liquid crystal state collapses to the liquid at 133 °C.

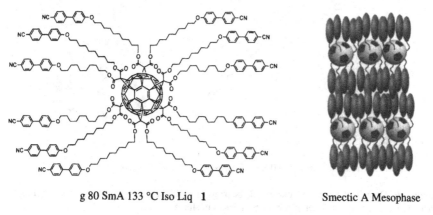

g 80 SmA 133 °C Iso Liq   **1**   Smectic A Mesophase

Figure 5. Supermolecular fullero-dendrimer **1** possessing cyanobiphenyl mesogenic units. The materials exhibits a lamellar smectic A phase.

Similarly material, **2**, shown in figure 6, is an octamer based the octasilsesquioxane core unit [6]. The material was found to exhibit smectic polymorphism with SmC and SmA phases being formed. The incorporation of cyanobiphenyl mesogenic moieties means that the material has interesting dielectric properties, and the presence of the smectic C phase indicates that with the introduction of chirality the material would also be ferroelectric and pyroelectric. The mesophase sequence of g −12.8 Cryst$_1$ 4.7 Cryst$_2$ 39.0 SmC 74.2 SmA 102.9 °C Iso Liq, demonstrates that the dendritic structure of the supermolecule is constrained to being rod-like, see the minimized structure in figure 6, and that the rod-like conformers tilt over at the smectic A to smectic C phase transition, as shown in figure 7. Thus, the structures of both the smectic A and the smectic C phases have alternating organic and inorganic and layers. As the inorganic and organic layers have differing refractive indices, the mesophase structures are essentially nano-structured birefringent slabs.

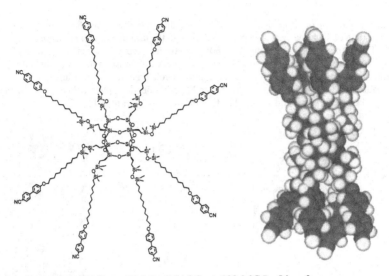

g −12.8 Cryst$_1$ 4.7 Cryst$_2$ 39.0 SmC 74.2 SmA 102.9 °C Iso Liq, **2**

Figure 6. Octasilsesquioxane octamer, **2**, bearing cyanobiphenyl moieties, and the minimized structure showing that the dendrimer is cylindrical in shape.

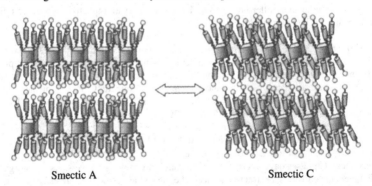

Smectic A   Smectic C

Figure 7. The structures for the SmA and SmC phases of supermolecular material **2**.

Apart from covalently binding mesogenic coatings to scaffolds such as fullerene and octasilsesquioxane, metal nano-particles can also be induced to form such hierarchal structures. For example, gold nano-particles coated with calamitic [7-10] and discotic [11-12] monolayers of mesogenic stabilizing ligands have been described. We have prepared a range of gold nano-particles coated with mesogenic thiols, eg **3**, and studied their behavior as dopants in nematic, smectic and cholesteric phases [13]. The stabilizing mesogenic cyanobiphenyl terminated thiols

266

chosen were designed to match perfectly the chemical nature of the liquid crystal solvent to be used as the host (ie E7 and 4-octyloxy-4'-cyanobiphenyl) in order to increase solubility and avoid the possibility of separation due to chemical incompatibility of the particle and the solvent. These nano-particles are highly soluble in the liquid crystal solvents studied without the need of sonication, giving dark brown solutions.

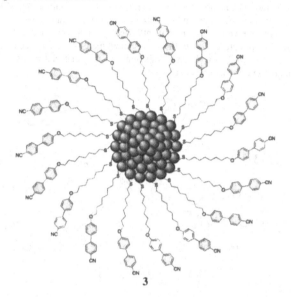

3

Figure 8. A gold nano-particle coated with cyanobiphenyl mesogenic units.

At room temperature they form a continuous nematic phase; however on heating near to the nematic to isotropic transition a strong phase separation occurs, demonstrating clearly domains of particle poor and particle rich areas which show different thermal behavior, eventually causing the separation of the particles from the isotropic liquid. The reverse is observed on cooling from the isotropic liquid into the continuous nematic phase. This behavior matches closely the theoretically predicted phase diagram of a mixture of hard particles in a nematic phase at high-weight fraction of mesogen [14].

Thus, the self-organization in two- and three-dimensional space offered by the liquid crystalline medium is an ideal vehicle to explore and control the organization of matter on the nanometer to the micrometer scale which is key to the emerging development of nano-technology.

Broken Symmetry

The introduction of chirality in liquid crystal systems can result in the exhibition of nonlinear properties such as ferro-, antiferro-, and pyro-electricity *etc*. Normally chirality is introduced into liquid crystal systems via the incorporation of stereogenic centers in the structures of the mesogens. We also used this approach in the design and synthesis of supermolecular materials with the result that new phase behavior was observed. This work also allowed us to examine how chirality is transmitted and propagated between nano-particles, with particular reference to the possibility of transference through chiral recognition surfaces.

The incorporation of lateral chiral mesogenic substituents onto a [60]fullerene scaffold led to interesting structural behavior and properties. Supermolecular material **4**, through bifurcation, possesses twelve mesogenic units attached symmetrically about the C_{60} core, thereby creating a spherical architecture. Due to the lateral attachment of the mesogenic units, a chiral nematic phase was found to be exhibited by this material. The material forms a glass at 47 °C, and a chiral nematic phase that is stable up to 103 °C.

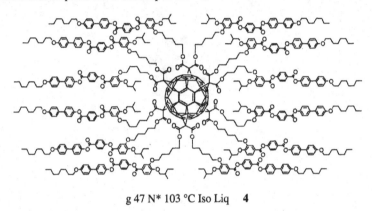

g 47 N* 103 °C Iso Liq   **4**

Figure 9. Supermolecular material **4** which exhibits a chiral nematic phase.

The material was annealed just below its clearing point, and after 24 hours, large areas of the preparation evolved to show fingerprint defects and the Grandjean plane texture of the chiral nematic phase. From the Grandjean plane texture the twist sense of the helical structure was found to be left-handed. The pitch was determined by measuring the number of pitch bands per unit length from the fingerprint texture. A value of 2.0 μm for the pitch length was obtained at room temperature. The value was found to be similar to those of the chiral mesogenic unit (1.7 μm) and the malonate precursor (1.9 μm). Thus the fullerene moiety is shielded very effectively among the laterally attached mesogens, without disturbing the helical supramolecular organization of the mesophase, which is shown in figure 10(a). Furthermore, as the mesogenic units are symmetrically distributed all over the fullerene sphere they effectively isolate it, thereby decreasing the possibility of aggregation of the C_{60} units, which is detrimental to mesophase formation.

It is also interesting to consider how the helical organization of **4** is created upon cooling from the liquid. As the C_{60} core of the material is spherical, and the mesogenic units are attached by relatively short methylene spacer units, it is not unreasonable to assume that, in the liquid phase, the mesogenic units are symmetrically disposed about the central core. Cooling into the chiral nematic phase, however, the helical organization was be expected to be a result of the

268

organized packing of the dendritic supermolecules, ie they are considered to be no longer spherical in shape. However, it was found that when the diameter of the C_{60} core is compared to the length of the mesogenic units, it is clear that flexible, random packing of the mesogenic units about the core in the liquid crystal state was not possible, and that the mesogens are required to be organized in their packing arrangements relative to one another, both on the surface of the dendrimer and between individual dendrimer molecules. For an individual dendrimer it was proposed that the direction of the mesogens would spiral around the C_{60} core to give poles at the top and bottom of the structure, as shown in figure 10(b). Thus the spherical dendrimer was projected to have a well-defined chiral surface, thereby resulting in the creation of a chiral nano-particle, *ie* a nano-molecular *"Boojum"*. When the chiral nano-particles pack together they were expected to do so through chiral surface recognition processes, resulting in the formation of a helical supramolecular structure.

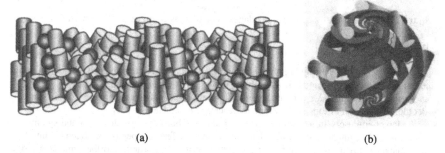

(a)   (b)

Figure 10: Proposed helical structure of supermolecule **4**. The fullerene units are shown as spheres embedded in the helical organization of the mesogenic units (a). The nano-particles is predicted to have a local spiraling of the mesogens thereby creating a molecular "Boojum".

Bifurcation

As shown above, bifurcation at the surface of the nano-particles can allow the space about the particle to be more efficiently filled. This is important for metal (eg Au) and inorganic nano-particles (eg Fe_3O_4) which have diameters in the range of 3 to 10 nm. One of the better ways to achieve this is through utilizing di- and tri-substituted phenyl dendritic groups as shown in figure 11. The attachments to the phenyl units may be aliphatic chains or mesogenic groups. Aliphatic chain attachment was found to lead to mesophases being formed primarily due to shape/topology considerations, whereas if mesogens are incorporated they will influence mesophase formation. In the first case there is a greater chance of forming cubic and columnar phases, whereas for the second lamellar phases tend to be found.

When X in figure 11 is replaces by an acid moiety the dendrons can be attached to various nano-particles composed of magnetic materials such as iron oxides and manganese clusters. The space filling natures of the dendrons can affect mesophase formation, and thereby long range periodic organization can be achieved on one hand, whereas fluid, liquid-like systems can be achieved on the other.

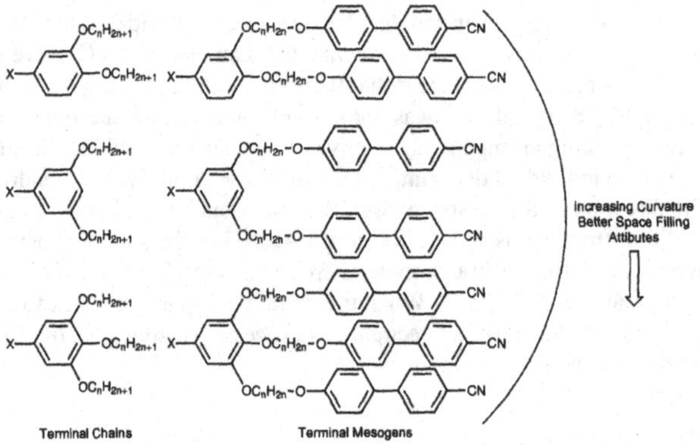

Figure 11. Bifurcation of terminal appendages to be used as coatings for nano-particles.

Hybrid Dependency

One of the more intriguing and challenging aspects in materials science is understanding the molecular recognition and self-assembling processes in materials with diversely functionalized faces or sides, which can yield supramolecular objects that may recognize and select left from right, or top from bottom, as described by de Gennes [15].

Janus liquid-crystalline molecular materials have been recently designed and synthesized in the form of segmented structures that contain two different types of mesogenic units [16]. Janus materials favor different types of mesophase structure, grafted onto the same scaffold, to create giant molecules that contain different hemispheres. The complementary materials **5** and **6**, based on a central scaffold made up of pentaerythritol and tris(hydroxymethyl)aminomethane units linked together, where one unit carries three cyanobiphenyl (CB) (smectic preferring), and the other three chiral phenyl benzoate (PB) (chiral nematic preferring) mesogenic moieties, or *vice-versa* were investigated [17-18].

On cooling **5** from the isotropic liquid a transition to a chiral nematic phase was found, followed by a second transition to a chiral smectic C* phase. Further cooling induced a glass transition at approximately -2.8 °C. It is also noteworthy that the transitions have extremely low ΔH values, suggesting that the system is relatively disordered and highly flexible. In addition, the formation of chiral mesophases by **5** means that the nematic phase is thermochromic and the smectic C* phase is ferroelectric and pyroelectric and exhibits electrostrictive properties. In contrast, compound **6** exhibits only a chiral nematic phase on cooling from the isotropic liquid. The only other thermal event present was a glass transition below room temperature at −7.9 °C.

Comparison of the phase behavior of compounds **5** and **6** shows clearly that the overall topology of the molecule with respect to the inner core plays a significant role in determining the type of mesophase formed, since in both cases the number of mesogens of each type and the core are the same and simply by placing them in different hemispheres changes the mesophase exhibited. For example, Janus material **7** [19] has one hemisphere with a dendron possessing terminal aliphatic chains which favors columnar and cubic phases, whereas the other hemisphere possesses a dendron with terminal rod-like mesogens that favors the formation of smectic phases. Combined together the Janus supermolecule exhibits either columnar or smectic phase.

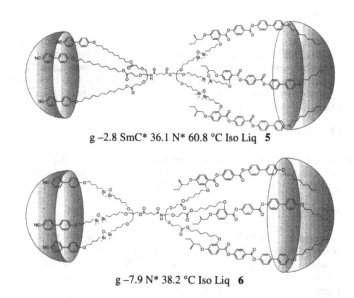

g −2.8 SmC* 36.1 N* 60.8 °C Iso Liq **5**

g −7.9 N* 38.2 °C Iso Liq **6**

Figure 12: Janus liquid crystal **5** and **6**.

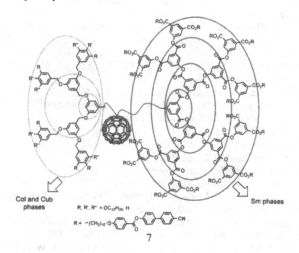

Col and Cub
phases

Sm phases

R, R', R" = OC₁₂H₂₅, H

R = −(CH₂)₁₈O⟨⟩CO₂⟨⟩⟨⟩CN

7

Figure 13. A Janus material that possesses a hemisphere favoring columnar and cubic phases, with the other favoring smectic phases

CONCLUSIONS

The manipulation of the structural fragments (mesogenic units, central scaffold, and linking units) in the molecular design of supermolecular systems potentially allows one to vary mesophase type and therefore the physical properties and potential applications of materials. Thus the molecular design of these systems is flexible and potentially capable of incorporating functional units, thereby allowing us to take some steps towards the molecular and functional complexity found in living systems [20-22].

ACKNOWLEDGMENTS

This work, as part of the European Science Foundation EUROCORES Program SONS was supported by the EC Sixth Framework Programme, under contract N. ERAS-CT-2003-980409 and by funds from our respective Research Councils.

REFERENCES

1. S. A. Ponomarenko, N. I. Boiko and V. Shibaev, *Polym. Sci. Ser. C,* **43**, 1; (2001).
2. V. Percec, M. Glodde, T. K. Bera, Y. Miura, I. Shiyanovskaya, K. D. Singer, V. S. K. Balagurusamy, P. A. Heiney, I. Schnell, A. Rapp, H. -W. Spiess, S. D. Hudson and H. Duank, *Nature,* **419**, 384 (2002).
3. J. -M. Fréchet, *Chem. Rev.,* **101**, 3819 (2001).
4. J. W. Goodby, G. H. Mehl, I. M. Saez, R. P. Tuffin, G. Mackenzie, R. Auzely-Velty, T. Benvegnu and D. Plusquellec, *Chem. Commun.,* 2057 (1998)
5. C. Tschierske, *J. Mater. Chem.,* **8**, 1485 (1998).
6. I. M. Saez, J. W. Goodby and R. M. Richardson, *Chem. Eur. J.,* **7**, 2758 (2001).
7. N. Kanayama, O. Tsutsumi, A. Kanazawa and T. Ikeda, *Chem. Commun.,* 2640 (2001).
8. I. In, Y. -W. Jun, Y. J. Kim and S. Y. Kim, *Chem. Commun.,* 800 (2005).
9. L. Cseh and G. Mehl, *J. Am. Chem. Soc.,* **128**, 13376 (2006).
10. L. Cseh and G. Mehl, *J. Mater. Chem.,* **17**, 311 (2007).
11. S. Kumar and V. Lakshminarayanan, *Chem. Commun.,* 1600 (2004).
12. M. Yamada, Z. Shen and M. Miyake, *Chem. Commun.,* 2569 (2006).
13. M. Draper, J. W. Goodby and I. M. Saez, *in press.*
14. M. Bates, *Liq. Cryst.,* **32**, 1525 (2005).
15. P. -G de Gennes, *Angew. Chem. Int. Ed. Engl.,* **31**, 842 (1992).
16. H. Stark, *Phys. Rev. E,* **66**, 032701 (2002).
17. I. M. Saez and J. W. Goodby, *Chem. Commun.,* 1726 (2003).
18. I. M. Saez and J. W. Goodby, *Chem., Eur. J.,* **9**, 4869 (2003).
19. J. Lenoble, S. Campidelli, N. Maringa, B. Donnio, D. Guillon, N. Yevlampieva and R. Deschenaux, *J. Am. Chem. Soc.,* **129**, 9941 (2007).
20. R. Deschenaux B. Donnio and D. Guillon, *New J. Chem.,* **31**, 1064 (2007).
21. I. M. Saez and J. W. Goodby, *Structure and Bonding,* **128**, 1 (2008).
22. J. W. Goodby, I. M. Saez, S. J. Cowling, V. Görtz, M. Draper, A. W. Hall, S. Sia, G. Cosquer, S. -E. Lee and E. P. Raynes, *Angew. Chem. Int. Ed.,* **47**, 2754 (2008).

Mater. Res. Soc. Symp. Proc. Vol. 1134 © 2009 Materials Research Society   1134-BB09-08

Microstructure and Morphology of Ba0.5Sr0.5TiO3-P(VDF-CTFE) Nanocomposites

Pei-xuan Wu, Xiaobing Shan, Jianli Song, Lin Zhang and Z.-Y. Cheng

Materials Research and Education Center, Auburn University, Auburn, Alabama 36849, USA

ABSTRACT

Ceramic-polymer 0-3 nanocomposites, in which nanosized $Ba_{0.5}Sr_{0.5}TiO_3$ (BST) powders were used as ceramic filler and P(VDF-CTFE) 88/12 mol% copolymer was used as matrix, were prepared and characterized over a concentration range from 0 to 40 vol.% of BST powders. It is found that the wetablity between the P(VDF-CTFE) copolymer and BST powders is poor, which results in poor uniformity in the as-cast composite film. Processes to modify the uniformity of the composites were investigated. It is found that the uniformity of the composite can be improved by hot pressing multilayers of the as-cast composite films. Functionalized silane was also used as agent to chemically treat the surface of BST powders. It is found that the silane coupling agent can significantly improve the connection between the BST powders and P(VDF-CTFE) copolymer matrix as reflected by a better uniformity observed in the as-cast composites.

INTRODUCTION

Materials with relatively high dielectric constant, low dielectric loss, low process temperature, and high flexibility are highly needed for applications ranging from electronic packaging to energy storage. Dielectric polymers are flexible and can be process at low temperature. More importantly, dielectric polymers can stand with a very high electric field (>500 MV/m), which makes most of current energy storage capacitors be made of polymers. However, it is well known that the polymers exhibit very low dielectric constants. On the other side, the inorganic ceramics, especially the ferroelectric-related ceramics, usually exhibit very high dielectric constant. Therefore, ceramic/polymer composites, especially ceramic-polymer 0-3 composites in which the ceramic powders are randomly filled in a polymer matrix, have been widely investigated. Those composite combines the good dielectric performance of ceramics and the high mechanical performance and chemical stability of polymers. It is expected that those composites are flexible and exhibit a relative high dielectric constant and can stand with high electric field. Based on the experimental results, it has been found that using nanosized ceramic powders can increase the loading of ceramic content and can improve the uniformity of those 0-3 composite [1].

To develop high performance 0-3 composites, $BaSrTiO_3$ (BST) powders have been utilized in the research of some composites due to the facts that the BST is a non-lead ferroelectric material and that the BST exhibits a high dielectric constant at room temperature. For example, in an investigation of composites in which the BST powders were filled in thermoplastic cyclic olefin copolymer (COC) [2], a dielectric constant of 13.9 with a dielectric loss of 0.014 was obtained at 100 kHz.

In this paper, the 0-3 composites using nanosized BST are reported. The microstructure and morphology of the composites were characterized. The relationship between the dielectric response and the microstructure in the nanocomposites is studied. The results indicate that the uniformity of the composites is very critical to achieve high performance composites. Two different methods to improve the uniformity of the nanocomposites were developed. One method is based on hot-press, and the other is based on the chemical modification of the BST powders.

EXPERIMENTS

In this study, P(VDF-CTFE) copolymer with a VDF/CTFE ratio of 88/12 mol%(VC88) was selected as the polymer matrix and $Ba_{0.5}Sr_{0.5}TiO_3$ (BST, from nGimat Co.) naonpowder was utilized as the ceramic filler. The BST-P(VDF-CTFE) 0-3 composites were prepared by a traditional solvent-cast method followed by a hot pressing (HP) techniques. To prepare the composite film, the P(VDF-CTFE) copolymer was first dissolved in N, N dimethylformamide (DMF) by using a magnetic stirring for 4 hours to make polymer solution. Then, the BST nano-powders were added into the polymer solution, which was then stirred for another 8 hours to get a relative uniform BST suspension. The BST suspension was cast onto a glass substrate at 70°C for 8 hours in an oven to form a solid composite film. Finally, the composite film was annealed at 140°C for 2 hours.

It is found that the uniformity of the as-cast composite film is poor. For example, a polymer-rich layer is found at the top of the as-cast composite film. To improve uniformity of the composites, the multiple layers of the as-cast composite film was stacked using two different configurations: in one configuration, the polymer-rich side of an as-cast composite film was placed to face the polymer-rich side of another as-cast composite film (named as PP configuration), in the other configuration, the polymer-rich side of an as-cast composite film was placed to face the non-polymer-rich side of another as-cast composite film (named as CP configuration). The stack was pressed using a hot press machine at 200°C for 30 seconds.

To improve the wetablity between BST powders and P(VDF-CTFE) polymer, a functional silane, 1H/1H/2H/2H-Perfluorooctyltrichlorosilane (97% from Alfa Aesar) was selected to chemically modify the surface of the BST powders as used in other research [3]. The silane coupling agent was dissolved in ethanol. The BST powders were first mixed with the silane-ethanol mixture. Ultrasonic dispersion was used to mix silane with the BST powders for 1 hour. Then, the BST powders with the silane-ethanol mixture were addied into the P(VDF-TrFE) solution in DMF. The mixture was stirred for another 8 hours. The mixture was cast onto a glass substrate and dried at 70°C for 8 hours in an oven to form the composite film. Finally, the composite film was annealed at 140°C for 2 hours.

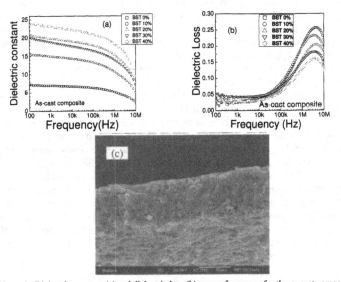

Figure 1. Dielectric constant (a) and dielectric loss (b) versus frequency for the as-cast composites at room temperature and the SEM picture of the nanocomposite with 40 vol.% BST.

To characterize the dielectric properties of the composites, gold thin film was sputtered onto the surfaces of the composite film to serve as electrodes. An Agilent 4294A impedance analyzer was used to measure the impedance of the composite samples at frequencies from 100 Hz to 10 MHz. The dielectric properties of the composites were calculated from the impedance using parallel plate capacitor mode. The microstructure and morphology of the composite samples were examined using a JEOL (JSM-7000F) Field Emission Scanning Microscope.

RESULTS AND DISCUSSION

The dielectric constant and loss vs. frequency of the as-cast composite film are shown in Figure 1(a) and 1(b). The microstructure of the composite observed using SEM is shown in Figure 1(c). Clearly, the uniformity of the as-cast composite film is poor as a polymer layer is observed at the top of the composite.

The results shown in Figure 2 are the dielectric response of the composite after the hot pressing (HP) process, where the composites with 40 vol.% BST were made from 4-layers of the as-cast composite film using both PP and PC configurations during HP process. It is found that the samples made using the PP configuration exhibit a high dielectric constant. Comparing the data shown in Figure 1, one can find that the HP process significantly enhances the dielectric response of the composites. For example, the dielectric constant at 1 kHz increase from 23 for the as-cast composite to 68 for the composites made from HP using PP configuration. As shown in Figure 2(c), the cross-section of the composites is pretty uniform and more compact than as-cast composite film. That is, the HP process significantly improves the uniformity of the composites.

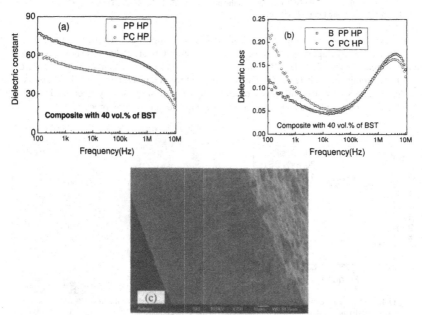

Figure 2. Dielectric constant (a) and dielectric loss (b) versus frequency as well as the SEM picture of the microstructure for the HP composites with 40 vol.% of BST. During the HP process, 4-layers of the as-cast composite film were used with both PP and PC configuration.

The data shown in Figure 3 are the dielectric response of the composites with 30 vol.% BST made using silane coupling agent. Clearly, the dielectric response of the composites is strongly dependent on the amount of silane used in the BST surface modification. Comparing the results shown in Figure 3 and Figure 1, one can find that the dielectric constant shown in Figure 3 is significantly higher than the composites shown in Figure 1 with the same content of BST powders. For example, the dielectric constant at 1 kHz increase from 19 to 49 (adding 260 μl silane), or the dielectric constant at 1 kHz increases from 19 to 29 (adding 600 μl silane) with a small dielectric loss 0.08.

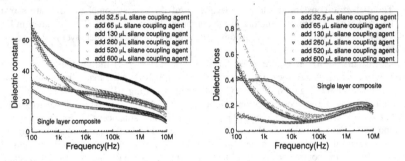

Figure 3. Dielectric constant (a) and dielectric loss versus frequency for the as-cast composites with 30 vol.% BST made from surface modified BST powders. In the process of surface modification, different amounts of the silane were assed into the BST powders.

When the hot-pressing technique is employed to the composites made from surface modified BST powders, a further imporvement in the dielectric response is observased as shown in Figure 4, where the composites with 30 vol. % BST are presented. A dielectric constant of 77 with a dielectric loss of 0.1 was obtained at 1 kHz in the nanocomposite at room temperature. The SEM observation clearly demonstrates that the uniformity of those nanocomposites is very good and composite interfacial layer is removed by using silane coupling agent and hot pressing process.

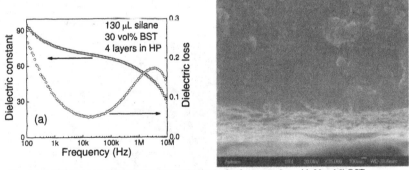

Figure 4. (a) Dielectric constant and loss versus frequency for the composites with 30 vol.% BST at room temperature, (b) the SEM picture of the cross-section of the same composite. The composites were made form the surface modified BST powders. The BST powders were modified using 130 μl silane.

276

CONCLUSIONS

Ceramic-polymer 0-3 nanocomposites with ceramic content from 0 to 40 vol.% are prepared with nanosized $Ba_{0.5}Sr_{0.5}TiO_3$ (BST) powders as filler and P(VDF-CTFE) copolymers as matrix. It is found that the wetablity between the polymer and ceramic powders in the as-cast composite is not good. The approaches to improve the uniformity of composites were investigated. It is found that the hot-press (HP) technique can significantly improve the uniformity of the composite, which results in a high dielectric constant in the composites. It is also found that the way/configuration to place the as-cast composite films together is very important. It is also found that when the silane as a coupling agent was used to modify the surface of the BST powders, the solution cast composites exhibit a higher dielectric constant and a much uniform microstructure.

REFERENCES

1. Dias, C. J., Igreja, R., Marat-Mendes, R., In´acio, P.,Marat-Mendes, J. N. and Das-Gupta, D. K: IEEE Trans. Dielectr. Electr. Insul., Vol. 11(1) (2004), p. 35-40.
2. T. Hu, J. Juuti, H. Jantunen, T. Vilkman: Journal of the European Ceramic Society, Vol. 27 (2007), p. 3997-4001.
3. Michael G. Todda and Frank G. Shi: Journal of applied physics, Vol. 94(7) (2003), p. 4551-4557.

AUTHOR INDEX

280

SUBJECT INDEX